71+10 New

Mathematics
Projects

Sumita Bose

V&S PUBLISHERS

Published by:

V&S PUBLISHERS

F-2/16, Ansari road, Daryaganj, New Delhi-110002
23240026, 23240027 • *Fax:* 011-23240028
Email: info@vspublishers.com • *Website:* www.vspublishers.com

Regional Office : Hyderabad
5-1-707/1, Brij Bhawan (Beside Central Bank of India Lane)
Bank Street, Koti, Hyderabad - 500 095
040-24737290
E-mail: vspublishershyd@gmail.com

Branch Office : Mumbai
Flat No. Ground Floor, Sonmegh Building
No. 51, Karel Wadi, Thakurdwar, Mumbai - 400 002
022-22098268
E-mail: vspublishersmum@gmail.com

Follow us on: 🇹 f in

For any assistance sms **VSPUB** to **56161**

All books available at **www.vspublishers.com**

Printed at : Param Offseters Okhla New Delhi-110020

Publisher's Note

V&S Publishers is glad to inform its esteemed readers that after a number of books under the **71series,** like *71+10 New Science Projects* (both in English & Hindi), *71 Science Experiments* (in English & Hindi), *71+10 New Science Projects Junior* (both in English & Hindi), *71+10 Magic Tricks for Children, 71 Arts & Crafts for School Children*, etc., it has now come up with this new, innovative and interesting book on Mathematics named **71+10 New Mathematics Projects**. The book has been written by a veteran school teacher, Sumita Bose having years of teaching experience in various reputed schools of Delhi and presently teaching in USA.

The prime objective of this book is to teach the students, particularly at the school level, the different basic concepts of Mathematics through simple, fun-filled projects with objects that are easily available in and around our natural surroundings. Since the book has been written by an experienced Mathematics teacher, all the projects are classroom tested explained in easy steps and divided into the four major categories of Mathematics, such as: *Arithmetic, Algebra, Geometry* and *Statistics*.

These step by step simple projects included in the book help to build up the confidence of the students to solve various kinds of problems in Mathematics and banish the phobia that many possess for this subject in their hearts and minds. These projects also help in clearing the basic concepts of Mathematics increasing their retention of knowledge, encouraging the students to understand the significance of Mathematics and Calculations in our day to day lives. All the Projects have their Solutions or Answers at the back of the book. In addition to the above, the projects also enhance the artistic and social skills of the students.

So, Dear Readers, especially the Young Ones, please go through these educative and fun-filled Projects thoroughly, as it will definitely prove to be a great learning experience for all of you!

<u>Introduction</u>

Mathematics is not limited to the four basic operations of **addition, subtraction, multiplication** and **division**. It is not a series of *assignments* and *worksheets*. It is an inseparable part of life.

Learning of mathematics best takes place through hands-on experiences. The concept building, fun-filled projects presented in this book are divided into four categories – **arithmetic, algebra, geometry** and **statistics.**

The step by step easy projects explained in this book help to banish the fear of mathematics lurking in some remote corner of the students' minds. The classroom tested projects of this book boosts up the self-confidence of students, increases retention of knowledge, encourages them to value mathematics as a part of life and develops proficiency in calculations. These projects include all the elements needed for a successful individual, partner or group learning experience. As an added bonus, the projects also enhance the artistic and social skills of the students.

General instructions for projects:

- ❏ Read each project completely before beginning it.
- ❏ Collect all the Materials Required for the particular project.
- ❏ Do not skip steps. Follow every step carefully.
- ❏ Note down your observations. If the results do not match with the results given in the book, read the instructions once again and redo the project from the beginning.

General precautions:

- ❏ Use a sharp pencil.
- ❏ Measure accurately.
- ❏ In the cutting and pasting projects, be careful not to use extra glue.
- ❏ In the paper folding projects, fold the paper uniformly.

<u>Acknowledgement</u>

"I hear and I forget.
I see and I remember.
I do and I understand."

– Confucius (Chinese Philosopher)

My heartfelt gratitude to my mother, husband and other family members for lighting my way with a soothing light of encouragement.

My heartiest wishes to all my students and contemporary teachers for keeping alight my passion for knowledge.

Finally, I am thankful to Mr. Sahil Gupta and his super team who toiled night and day to make this book a success.

Contents

Arithmetic Projects

Distributive Property

Aim

Materials Required

Prerequisite Knowledge

$$a \times (\quad + c) = \quad + ac$$

Procedure

Express the individual addends as a product of two numbers.

grid.

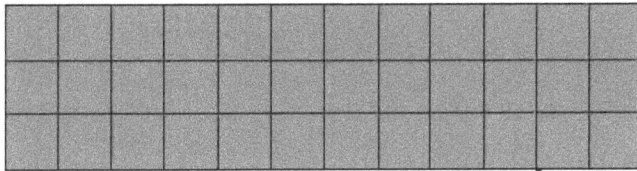

36 = 3 × 12

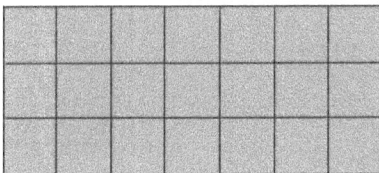

21 = 3 x 7

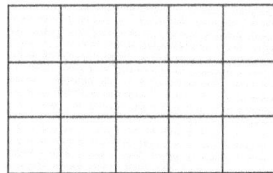

15 = 3 x 5

Observation

Number		
36		
21 + 15 (Sum of addends)		
factor)		

Result

Integers

Aim

Materials Required

Prerequisite Knowledge

Integer

Zero pair

integers.

Procedure

and the other

as

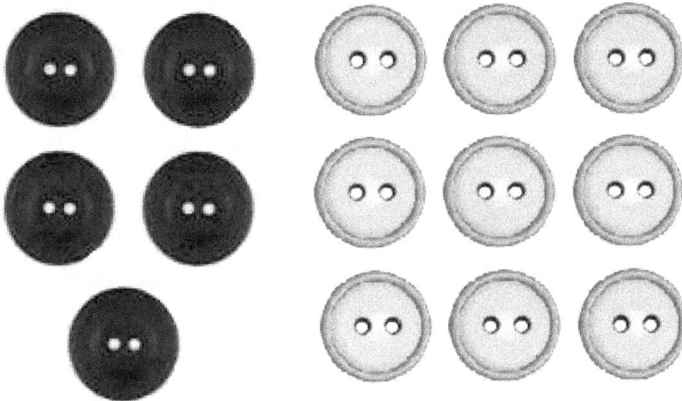

Observation

Integers	First integer	Second integer	Sum
+4 and +6			

Result

Prime and Composite Numbers

Aim

Materials Required

Prerequisite Knowledge

prime

Procedure

composite

represents a

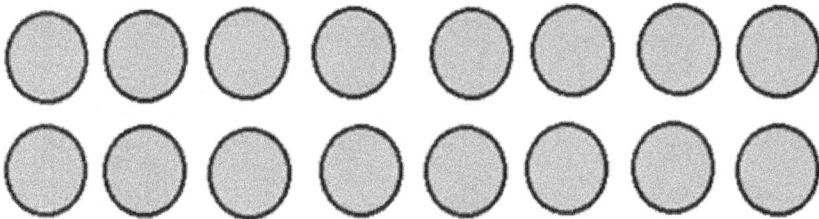

Observation

	Arrangement of	Dimensions	

Result

Factors and Multiples

Aim

Materials Required

Prerequisite Knowledge
 –
 –

factor of that number.

Procedure

Observation

Square grid for factors of 16	Dimensions of the grid

	Dimensions of the grid

Result

Prime Factorisation

Aim

Materials Required

Prerequisite Knowledge

prime

Procedure

Draw, colour and cut out two trees,

thin cardboard.
On the drawing sheet, paste a tree on top and write 30 on it.

numbers (except 1).

apple and if it is composite, write it on

the tree and draw two lines from the tree to connect the fruit and the

Now leave the apple number and further express the number on

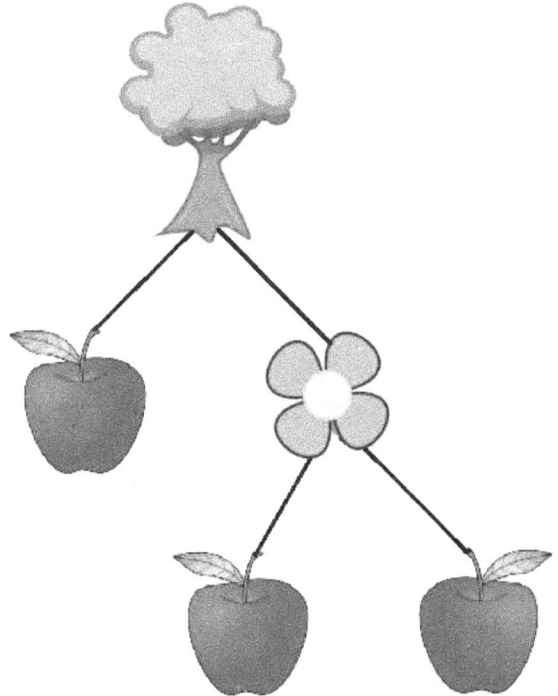

numbers.

. The product

Observation

The prime factors of 30 are _____.
The prime factors of 54 are _____.

Result

Highest Common Factor (HCF)

Aim

Materials Required

Prerequisite Knowledge

–

factor of that number.

Procedure

exceed the length of the bigger rectangle.

repeat the previous step.

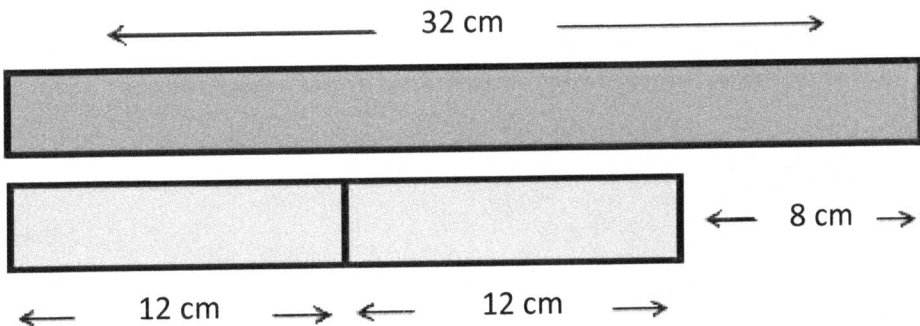

Two rectangles of length 12 cm each can be placed on the big rectangle

Observation

Lengths of rectangles		Number of rectangles of longer length	Number of rectangles of shorter length

Result

Improper and Mixed Fractions

Aim

Materials Required

Paper plates (six), a pair of scissors.

Prerequisite Knowledge

proper

denominators are called

mixed

Procedure

and the third pair into

represent 3/2, 9/4 and 10/3.

3/2 represents three halves, 9/4 represents

10/3 represents ten

table.

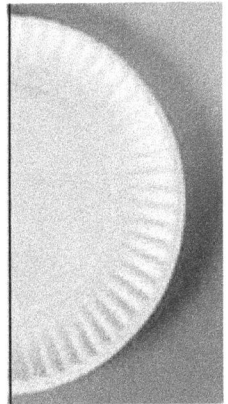

Observation

	Meaning of the		

Result

71 + 10 New Mathematics Projects

Comparing Fractions

Aim

Materials Required

Prerequisite Knowledge

less than 1.

Procedure

and 3/6, 5/6.

$\frac{1}{4}$	$\frac{1}{4}$	$\frac{1}{4}$

$\frac{1}{11}$	$\frac{1}{11}$	$\frac{1}{11}$	$\frac{1}{11}$	$\frac{1}{11}$	$\frac{1}{11}$	$\frac{1}{11}$

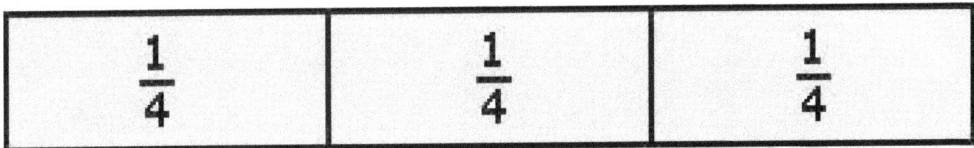

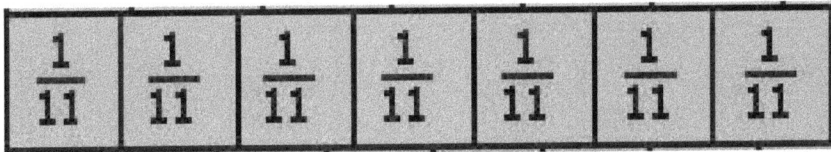

$$\frac{3}{4} \quad \frac{7}{11}$$

Observation

3/4, 7/11		
2/7, 5/12		
7/8, 6/8		
3/10, 1/3		
1/2, 4/9		
3/6, 5/6		

Result

Addition of Fractions

Aim

Materials Required

Prerequisite Knowledge

–

–

Procedure

common denominator.

one. One and the remaining

$$\frac{7}{8} \quad \frac{1}{4}$$

= $\boxed{\frac{1}{8}}\boxed{\frac{1}{8}}\boxed{\frac{1}{8}}\boxed{\frac{1}{8}}\boxed{\frac{1}{8}}\boxed{\frac{1}{8}}\boxed{\frac{1}{8}}$ + $\boxed{\frac{1}{4}}$

(a) 7/8 + 1/4   (b) 2/6 + 1/2

= $\boxed{\frac{1}{8}}\boxed{\frac{1}{8}}\boxed{\frac{1}{8}}\boxed{\frac{1}{8}}\boxed{\frac{1}{8}}\boxed{\frac{1}{8}}\boxed{\frac{1}{8}}$ + $\boxed{\frac{1}{8}}\boxed{\frac{1}{8}}$

= $\boxed{\frac{1}{8}}\boxed{\frac{1}{8}}\boxed{\frac{1}{8}}\boxed{\frac{1}{8}}\boxed{\frac{1}{8}}\boxed{\frac{1}{8}}\boxed{\frac{1}{8}}\boxed{\frac{1}{8}}\boxed{\frac{1}{8}}$

= $\boxed{1}$ $\boxed{\frac{1}{8}}$

Sum of 7/8 and 1/4

Observation

Problem	Addends	

Result

Multiplication of Fractions

Aim

Materials Required

Prerequisite Knowledge

Procedure

$$\frac{1}{5} \text{ of } \frac{3}{4} \quad \frac{1}{5} \times \frac{3}{4} \quad \frac{3}{20}$$

Observation

Problem	(Red squares)	(Blue squares)	Product
1/5 of 3/4			
5/6 of 7/8			

Result

Decimal Representation

Aim
To represent the decimal numbers on various *grid papers*.

Materials Required

Prerequisite Knowledge
–   *decimal*

an integer part and

Procedure

grid, hundredths grid and

drawing sheet).

rectangle represents 0.1,
in hundredths grid, each

in thousandths grid, each
small rectangle represents
0.001.

grids to represent 0.8,
0.009, 0.36, 1.2, 2.29, and
3.023.

One whole

Tenths grid

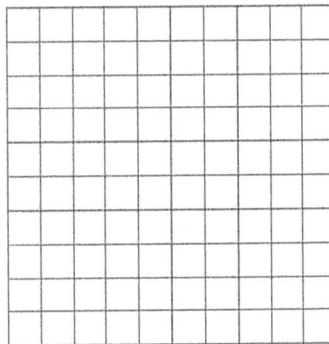

Hundredths grid

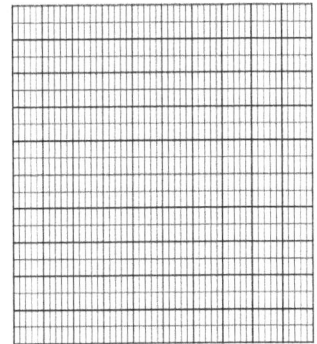

Thousandths grid

Observation

Number	
0.8	
0.009	
0.36	
1.2	
2.29	
3.023	

Result

Project - 12

Decimal Equivalent

Aim

grid papers.

Materials Required

Prerequisite Knowledge

–
decimal

Procedure

Draw or cut out the tenths grid, hundreds grid and thousandths grid as in the previous

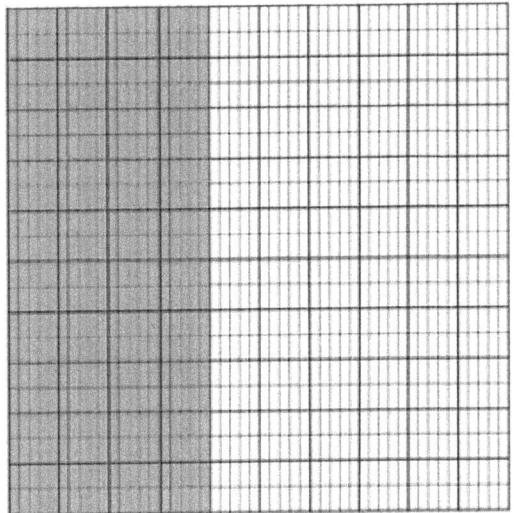

Observation

Decimal number	
0.4	
0.40	
0.400	
0.6	
0.60	
0.600	

Result

Addition of Decimals

Aim
To add the decimals using various *square grids*.

Materials Required

Prerequisite Knowledge

in a decimal number does not change the value

of the decimal number.

Procedure

hundredths grid.

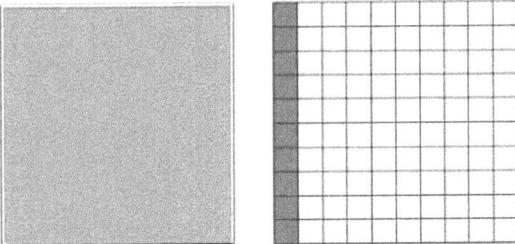

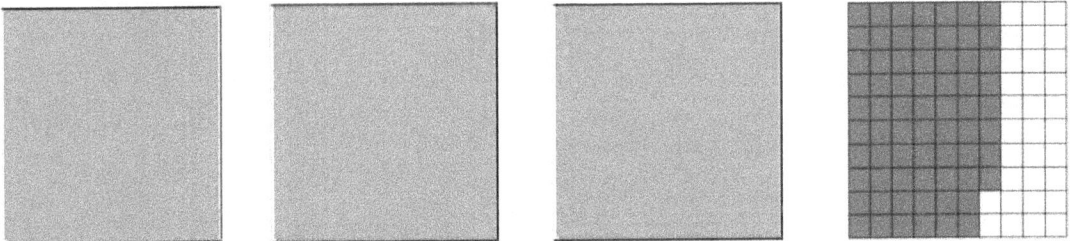

Observation

Problem	
Second addend	
Total number of whole and hundredths	
Sum	

Result

Measurement

Aim

paper skeleton and *actual measurement*.

Materials Required

Prerequisite Knowledge

–

Procedure

friend.

and feet.

readings.

on the top.

actual measurements.

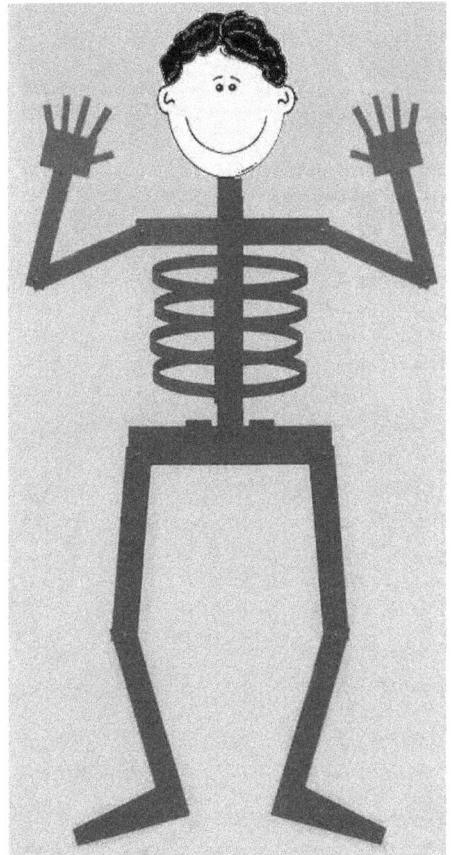

Observation

		Actual measurement
Palm		
Hand(shoulder to elbow + elbow to wrist)		
Shoulder		

Result

71 + 10 New Mathematics Projects

Exponents

Aim

paper folding.

Materials Required

Ruler, pencil, drawing sheet, a pair of scissors.

Prerequisite Knowledge

–

exponent m^a, *m* is the   and *a* is the *exponent.*

Procedure

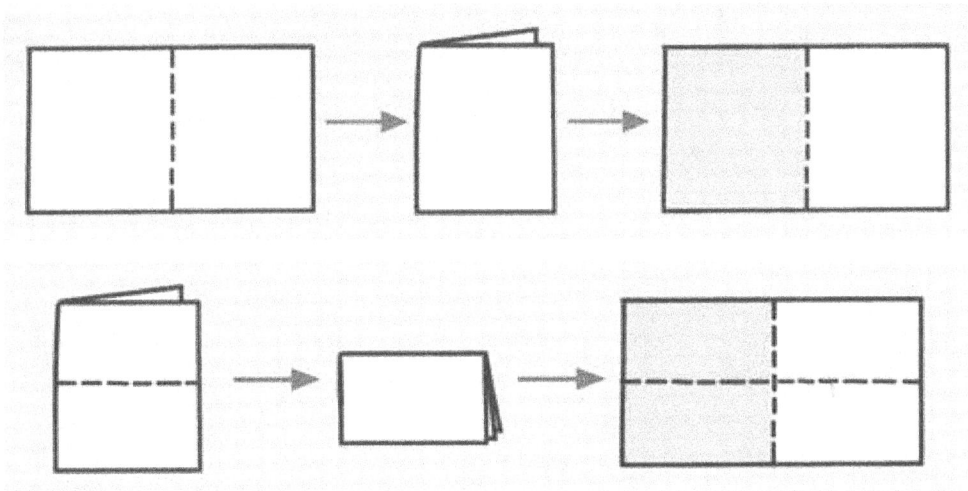

Observation

Number of folds		

Result

Project - 16

Square Root

Aim
To represent a *square root*   *grid paper.*

Materials Required

Prerequisite Knowledge
— *The square root*

Procedure

also draw them).

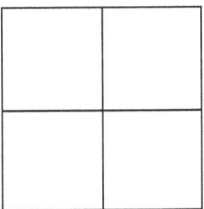

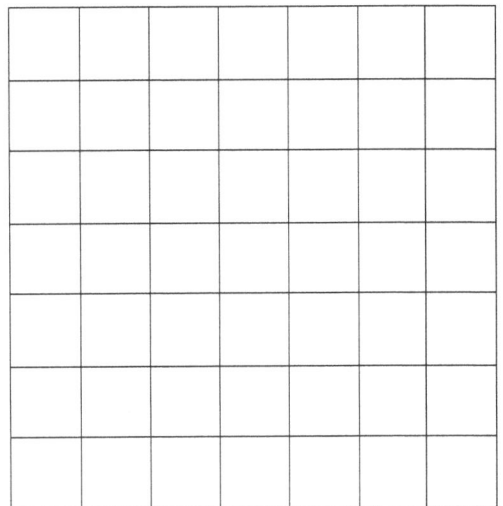

Observation

	Area = Total number of unit squares	squares on each side

Result

71 + 10 New Mathematics Projects

Percentage

Aim

, decimals and *percentage*.

Materials Required

Prerequisite Knowledge

– The word, percent or per cent means out of hundred.

Procedure

drawing sheet).

Draw the given designs (one design on each grid).

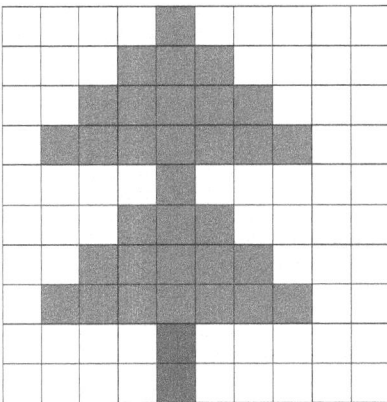

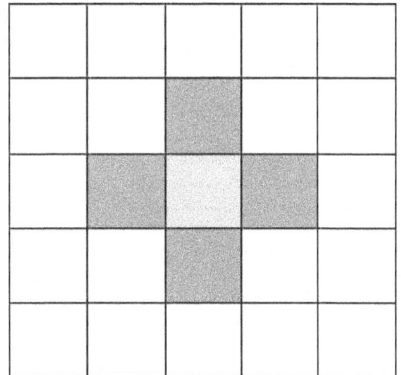

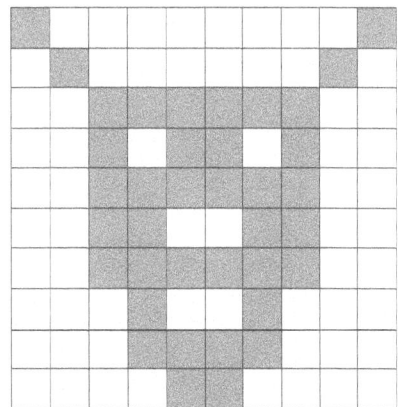

Observation

Design	Total number of squares	Number of squares coloured		Decimal	Percent

Result

71 + 10 New Mathematics Projects

Ratio

Aim

Materials Required

Prerequisite Knowledge
  -   *a*
and
Procedure

table.

wrist.

height.

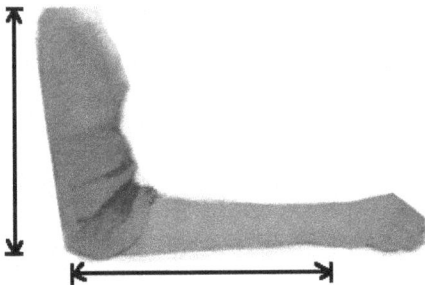

Note down the readings.

the elbow and the wrist.

Observation

Lengths/Height	Measurement in cm

$$\frac{\text{Length between your elbow and shoulder}}{\text{Length betwwn your elbow and wrist}} \quad \underline{\hspace{2cm}}$$

$$\frac{\text{Length between the two finger tips}}{\text{Your height}} \quad \underline{\hspace{2cm}}$$

Result

Proportion

Aim

Materials Required
Ruler, measuring tape

Prerequisite Knowledge
-

are

Procedure

of a tree.
Stand next to the tree.

Note down the various data in

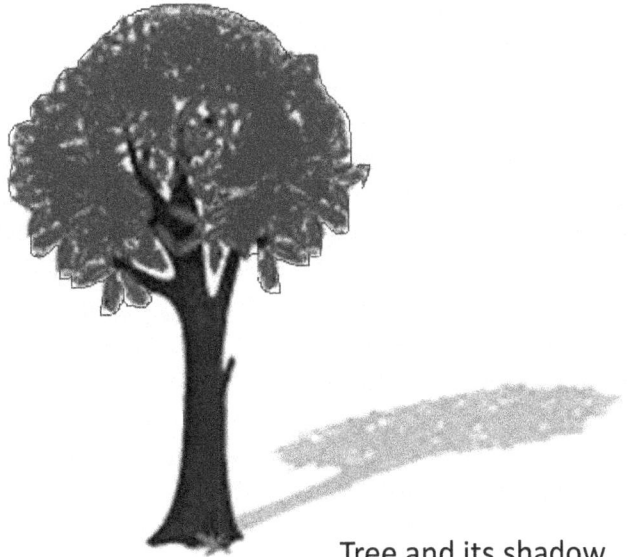

Tree and its shadow

$$\frac{\text{Height of a tree}}{\text{Length of a tree's shadow}} \quad \frac{\text{Your height}}{\text{Length of your shadow}}$$

Observation

Length/Height	

Result

Variation

Aim
To model _____ using *right angled triangles.*

Materials Required

Prerequisite Knowledge

x and y _____ $x : y$ remains constant.

x and _____ if the product xy remains constant.

Procedure

Place them one over the other matching one of the acute angles.

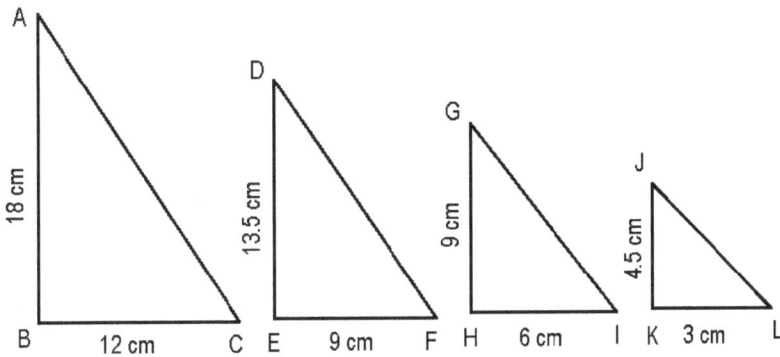

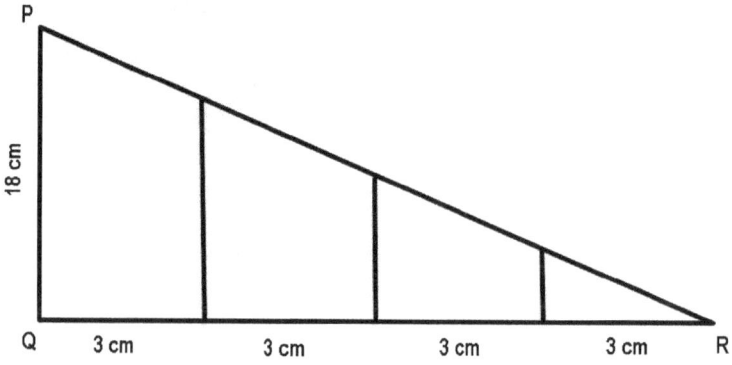

Observation

Name of the triangle	Perpendicular (in cm)	(in cm)	Perpendicular / Base

Result

Magic Square

Aim

To construct a

Materials Required

Prerequisite Knowledge

–   *equal rows and columns* in such a

Procedure

Draw a

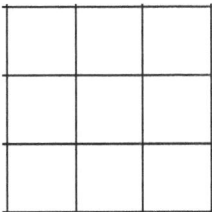
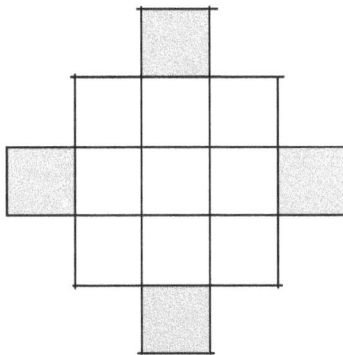
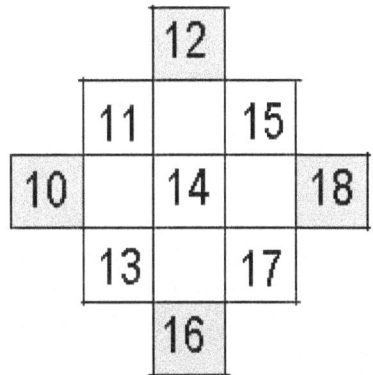

Magic Square   51

Observation

Sum of rows =_____

Sum of columns = _____

Sum of diagonals = _____

Result

Coding Decoding

Aim
To use   to   .

Materials Required
Drawing sheet, 85 small

Prerequisite Knowledge
Braille –

Procedure

them).

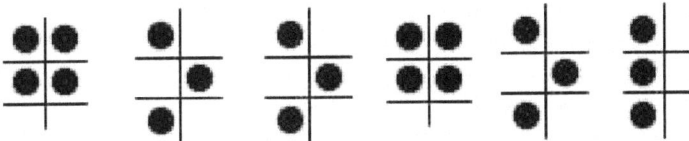

and   .

a b c d e f g h i j

k l m n o p q r s t

u v w x y z

Observation

Result

Candle Clock

Aim

Materials Required

watch, ruler.

Prerequisite Knowledge
The wax burns at a constant rate.

Procedure

burnt out candle.

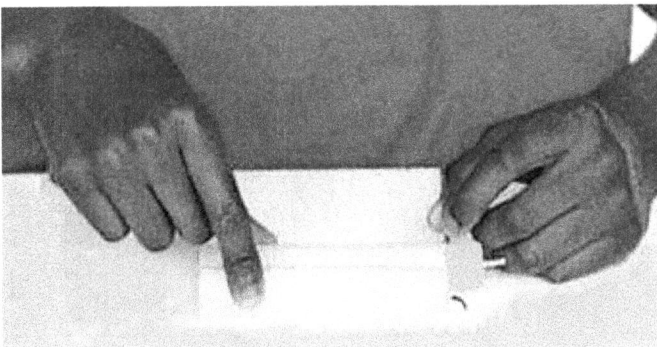

Observation
The length of the new candle = _____cm.

Result

Algebra
Projects

Representing Polynomials

Aim

and *polynomial.*

Materials Required

Prerequisite Knowledge

-   *monomial.*

–

called a   .

-   and *constants*
                                    and   *polynomial.*

Procedure

                              and the other as
x^2 (since each side is of length x units), the rectangle, x (since the length is x units and

$x, 3x$ x^2 $x^2, 3x^2$ x

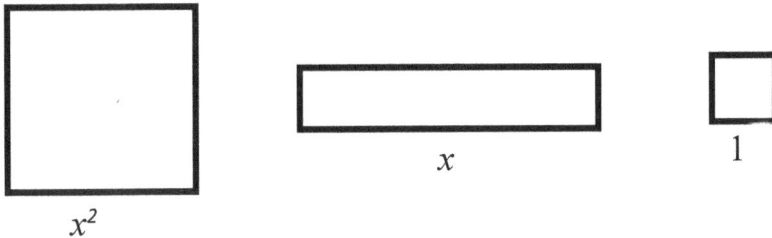

x^2

x

1

Fig:

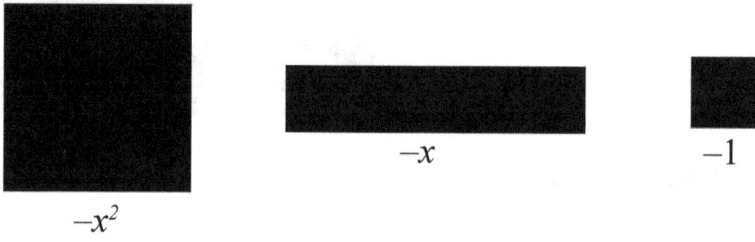

$-x^2$

$-x$

-1

Fig:

Observation

Algebraic		

Result

Representing $(3a)^2$ and $3a^2$

Aim

a^2 and $(3a)^2$.

Materials Required

Prerequisite Knowledge

$: a^0 = 1$

$^m)^n = a^{mn}$

$a^{-n} = \dfrac{1}{a^n}$

$^m \cdot a^n = a^{m+n}$

$\dfrac{a^m}{a^n} = a^{m-n}$

Procedure

cardboard.

interpret the result.

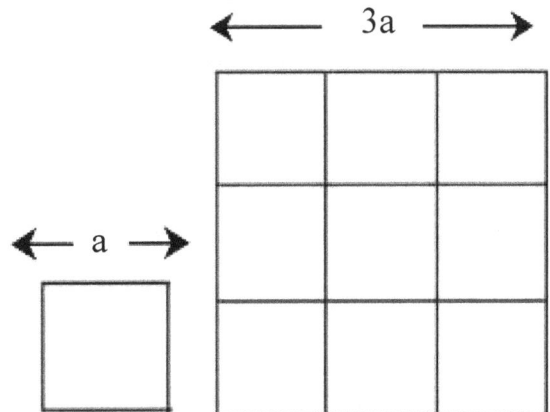

Fig: *a* and *3a*

Observation

Algebraic		
a^2		
$3a^2$		
$(3a)^2$		

Result

Adding Polynomials

Aim
To use   *polynomials.*

Materials Required

Prerequisite Knowledge
Like terms –

     -

Procedure

Represent the algebraic expressions $4x^2$ ^{2}

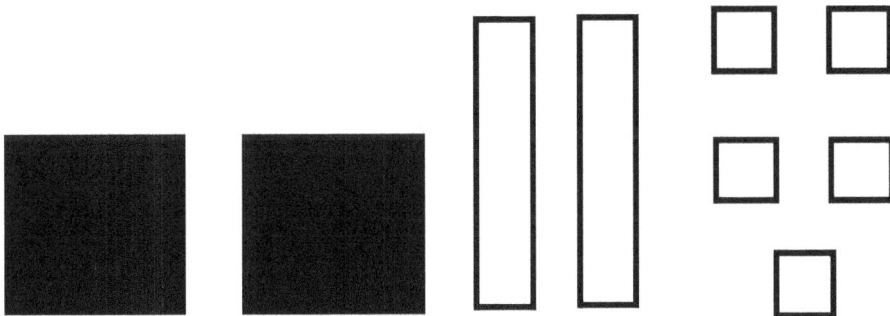

Fig: $^2 + 2x + 5$

Observation

$4x^2$	
$^2 + 2x + 5$	
$(4x^2 \qquad ^2 + 2x + 5)$	

Result

Distributive Property

Aim

Materials Required

Prerequisite Knowledge

$a \times (b + c) = a \times b + a \times c$ where a, b and c are *expressions.*

Procedure

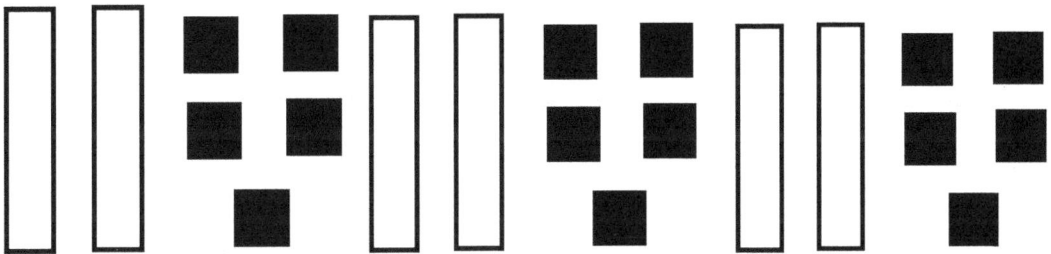

Fig: *x*

Observation

Result

Algebra Identity $(a + b)^2 = a^2 + 2ab + b^2$

Aim

To prove $(a + \quad)^2 = a^2 + 2 \quad + \quad ^2$

Materials Required

Prerequisite Knowledge

–

$= \text{Side} \times \text{Side}$

$=$

Procedure

Draw two perpendicular lines from the end

from a thin cardboard and cover them with a

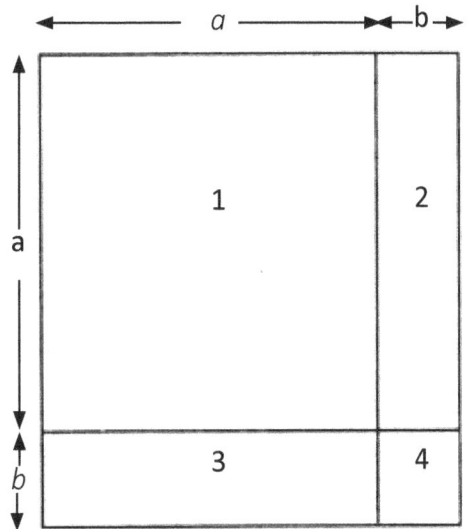

Fig:

Algebra Idenity $(a + b)^2 = a^2 + 2ab + b^2$

Observation

Side of the	Area of the	on the big square
a		
Rectangle of sides, and a		
Rectangle of sides, a and		
$a +$)		

Result

Algebra Identity $(a - b)^2 = a^2 - 2ab + b^2$

Aim

To prove $(a \quad)^2 = a^2 \qquad + \quad ^2$.

Materials Required

Prerequisite Knowledge

–

$\quad$ = Side × Side

$\qquad$ =

Procedure

a units.

sides, draw two perpendicular lines

side '
side '
'

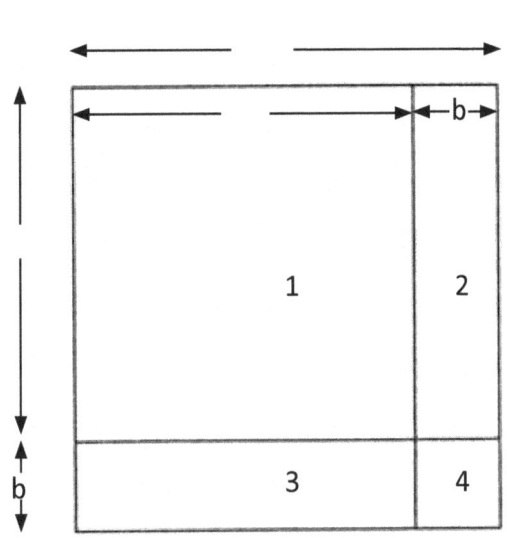

Fig:

Observation

Side of the geometrical	Area of the geometrical	big square
a		
Rectangle of sides, and		
Rectangle of sides, a and		
a)		

Result

Difference of Squares

Aim

To prove a^2 $^2 = (a$

Materials Required

Prerequisite Knowledge

= Side × Side

=

Procedure

paper.

Rearrange the rectangles to form a big rectangle (turn the second rectangle, join side

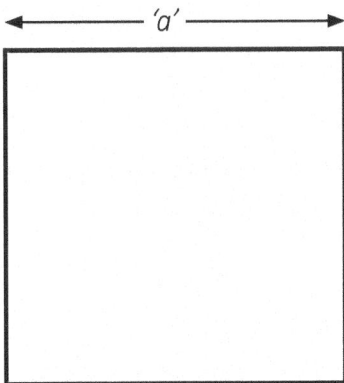

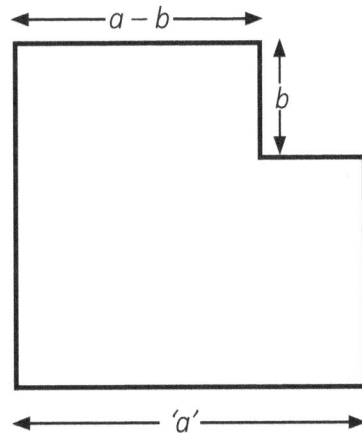

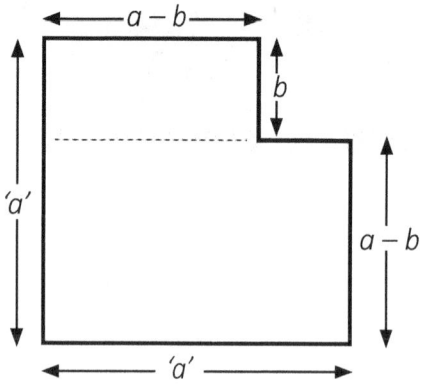

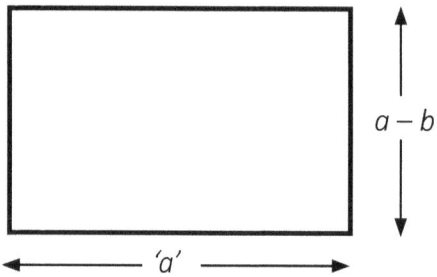

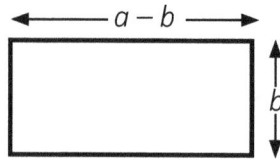

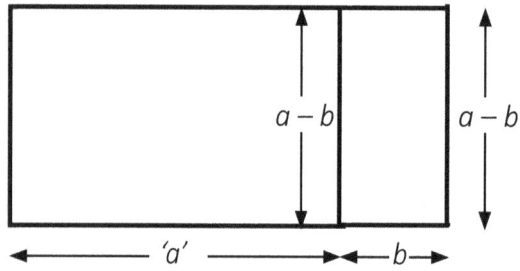

Fig:

Observation

a	
b	

Result

Factorisation of Algebraic Expression

Aim
To *geometrically* determine the *factors* of $x^2 + 7x$ and *middle term*

Materials Required

Prerequisite Knowledge

–

= Side × Side

=

Procedure

x units.

Draw three rectangles of sides x x

new rectangle.

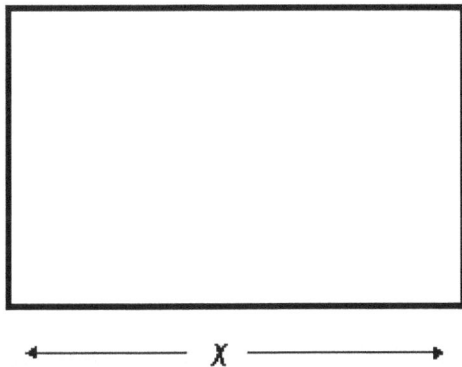

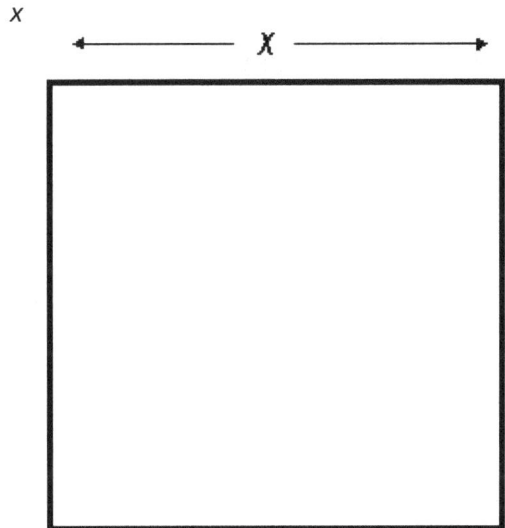

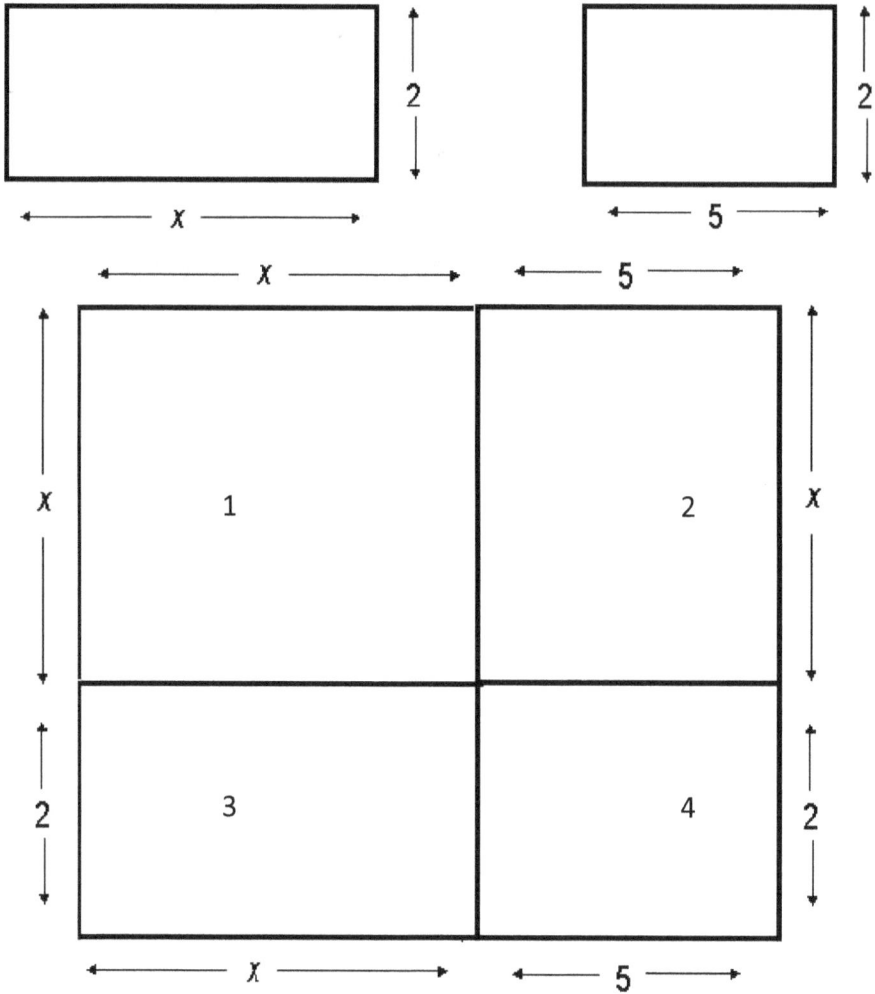

Fig: $x^2 + 7x + 10$

Observation

	Area of the	on the big square
Rectangle of sides, x and 5		
Rectangle of sides, x and 2		
Rectangle of sides, 5 and 2		
Rectangle of sides, (x + 5) and (x + 2)		

Result

Addition of Linear Equation

Aim

cups and counters.

Materials Required

Prerequisite Knowledge

having the highest power of the variable as one is

Procedure

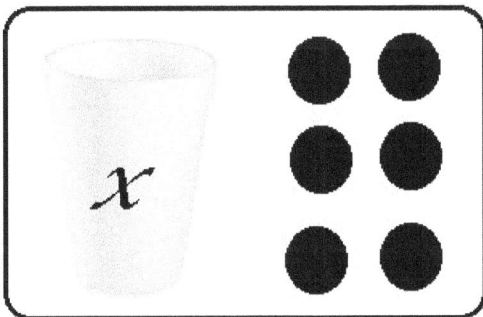 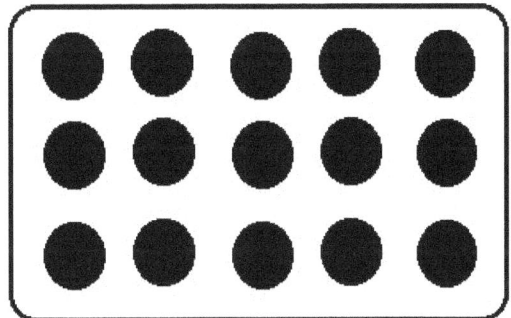

x + 6 = 15

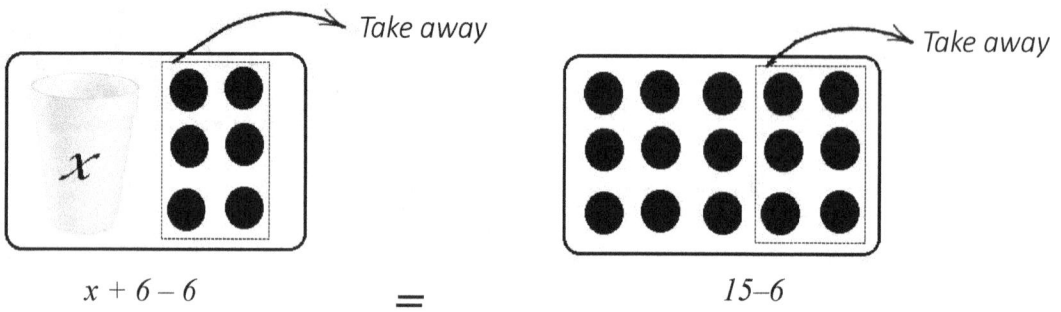

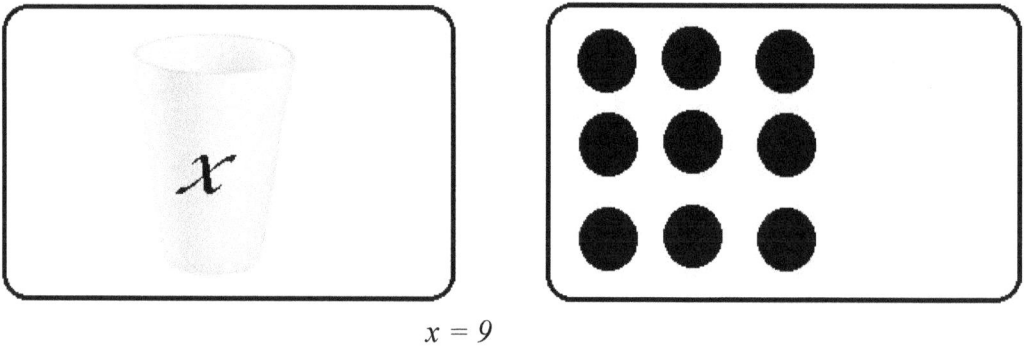

$x + 6 - 6$ = $15 - 6$

$x = 9$

Fig:

cups and counters.

Observation

		Right hand side of	
x + 6 = 15			

Result

Project - 33

Subtraction of Linear Equation

Aim

x .

Materials Required

Prerequisite Knowledge

–

a   .

-

Procedure

x

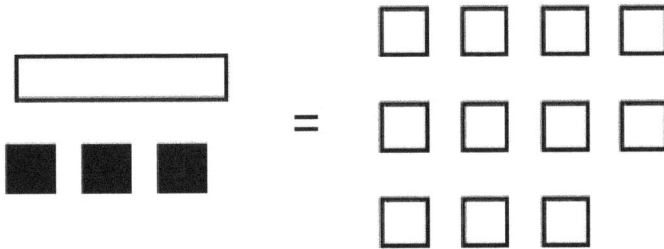

Fig:

Observation

	Algebraic/Numeric	
Right hand side		
pair		

Result

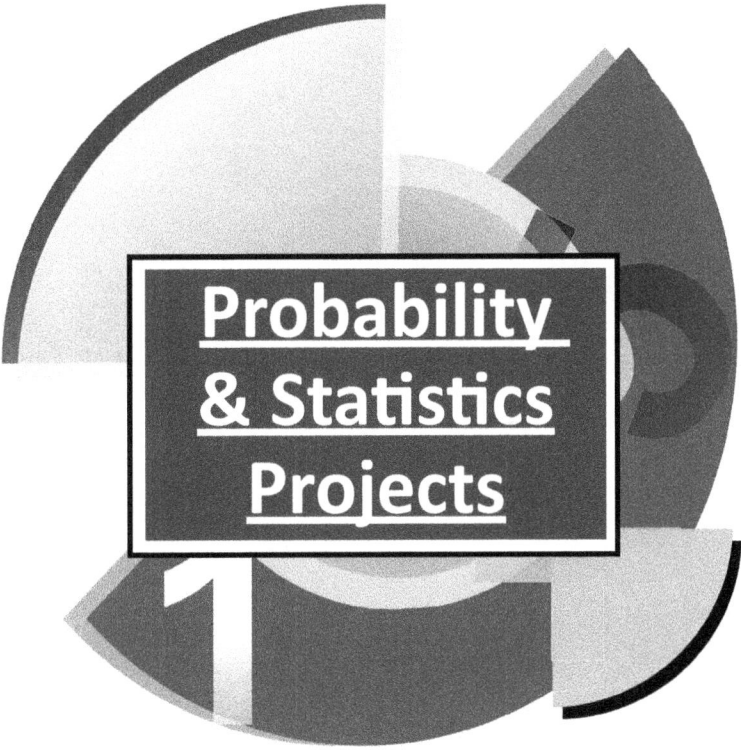

Probability & Statistics Projects

1

Coordinates

Aim

To draw the correct shape at the given coordinate points on a coordinate grid in order to obtain

Materials Required

Prerequisite Knowledge

–   *axes*

x-axis   *y-axis.*

–

) *a* represents the *x*   represents the *y*

coordinate.

Procedure

x   *y*

Draw appropriate shapes at the given coordinates.

■

◤

◢

◥

◤

I 5, I 7

Fig:

Observation

Result

Enlargement of a Picture

Aim

To use a coordinate grid to enlarge a drawing.

Materials Required

Prerequisite Knowledge

–

x-axis   *y-axis.*

–

ordered pair.   ), *a* represents the *x*   represents the *y*
coordinate.

Procedure

coordinate grid on a drawing sheet.

(2,2), (2,3), (2,5), (5,3), (2,3) , (2,2), (5,2), (4,1), (1,1).

from (1,1) and ending at (1,1).

Plot the new coordinates in a 15 × 15 grid and join the coordinate points in order

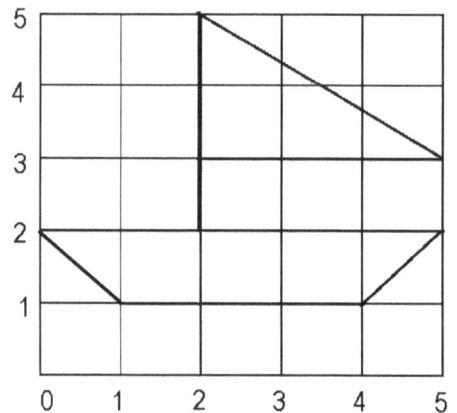

Fig: 5 × 5 grid showing a boat

Observation

Result

71 + 10 New Mathematics Projects

String Art

Aim
To draw four curves using straight lines and get a

Materials Required

Prerequisite Knowledge

axes.

x-axis   *y-axis.*

Procedure

sheet.
Paste the drawing sheet
on a thin cardboard.

thread in a needle.

Keep the needle below

one is chosen, insert the

of number one on the
y-axis
it down from the one on
the *x-axis*

the *y-axis* and put it down
from two on *x-axis*.

					1 ·						
					2 ·						
					3 ·						
					4 ·						
					5 ·						
					6 ·						
6	5	4	3	2	1	1	2	3	4	5	6
					1 ·						
					2 ·						
					3 ·						
					4 ·						
					5 ·						
					6 ·						

Fig :

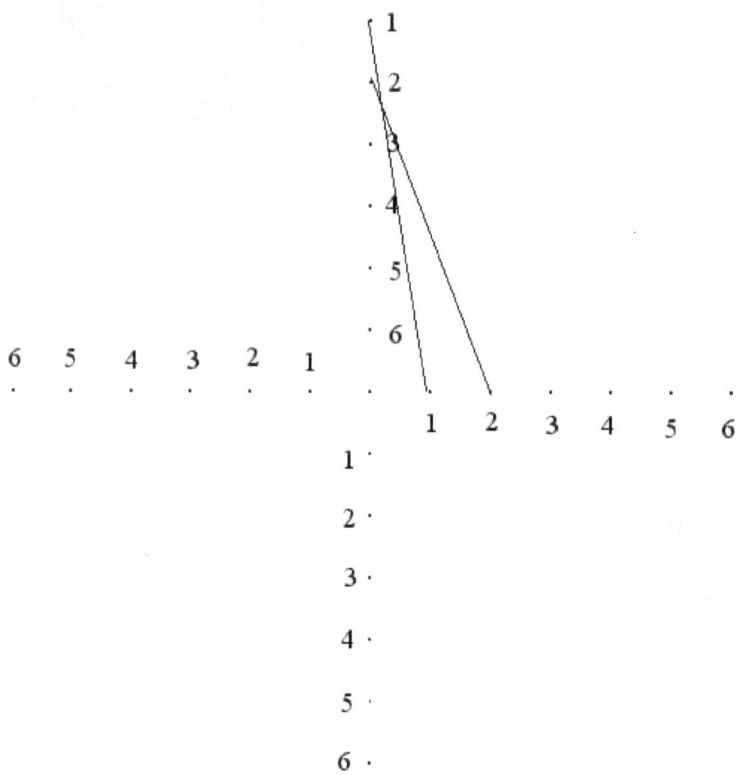

Fig : String art showing straight lines

Observation

Result

Line Graph

Aim

To measure the pulse rate during rest and exercise and draw its line graph.

Materials Required

Prerequisite Knowledge

pulse rate.

Procedure

minute.

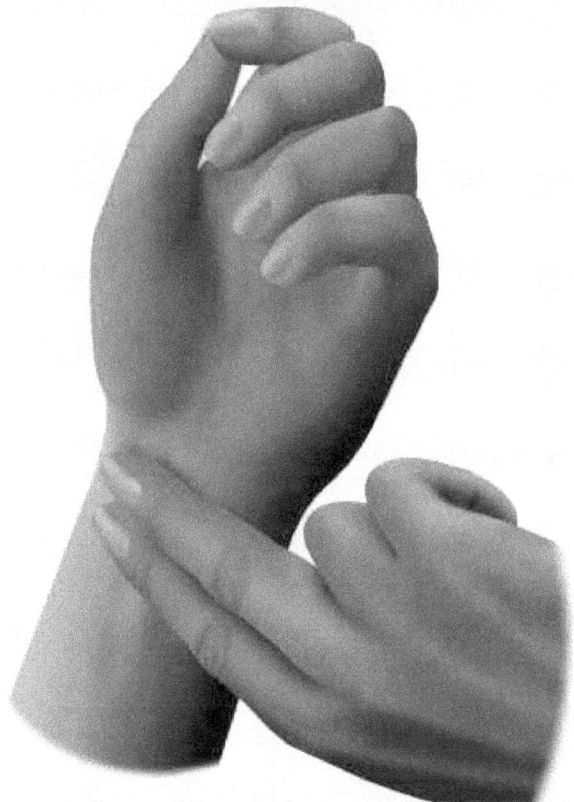

again measure the pulse rate.

and draw a line graph in the following format.

Fig:

Observation

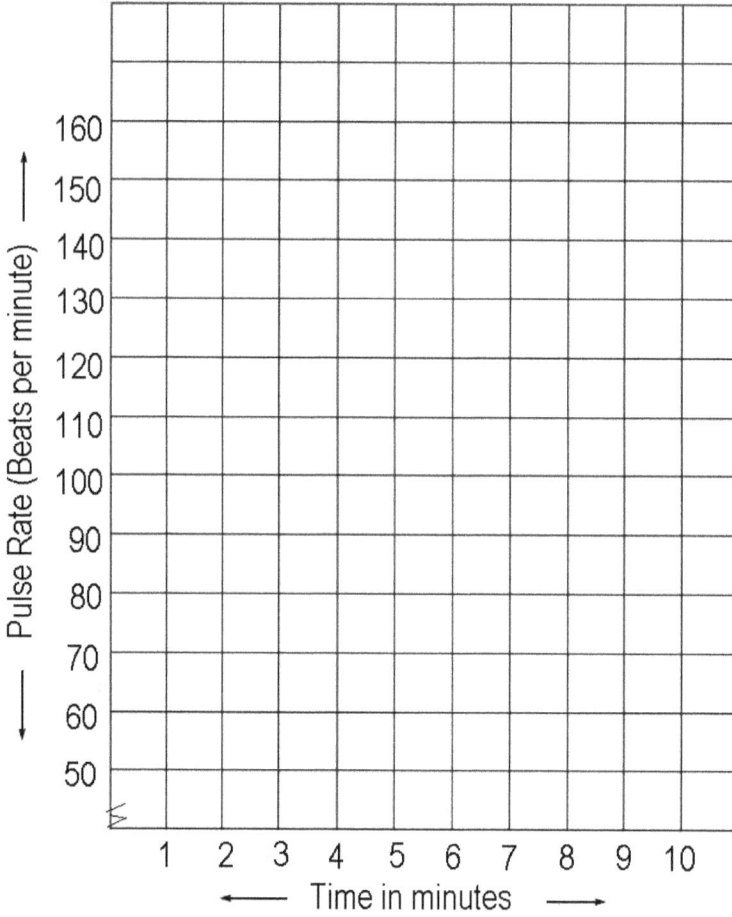

Fig: Line graph format

Pulse rate during exercise = _____

Result

Bar Graph

Aim
To collect *data* and draw a   *gems chocolate*.

Materials Required

of scissors.

Prerequisite Knowledge

–

Procedure

gems of each colour as per the scale chosen.

on the *y-axis* and the *colours* on the *x-axis.*

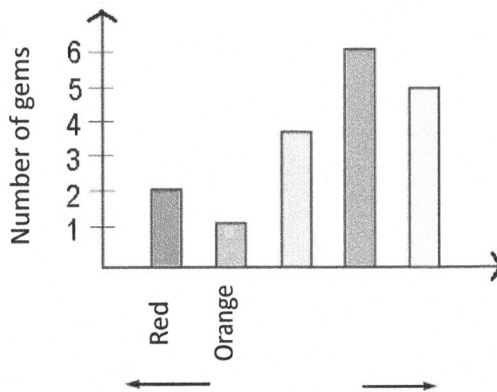

Fig:

Observation

Result

71 + 10 New Mathematics Projects

Double Bar Graph

Aim
To draw a double bar graph and *analyse* the graph.

Materials Required

Prerequisite Knowledge
—

Procedure

lengths according to the number of

as per the scale chosen.

number of hours on the *y-axis* the
x-axis.

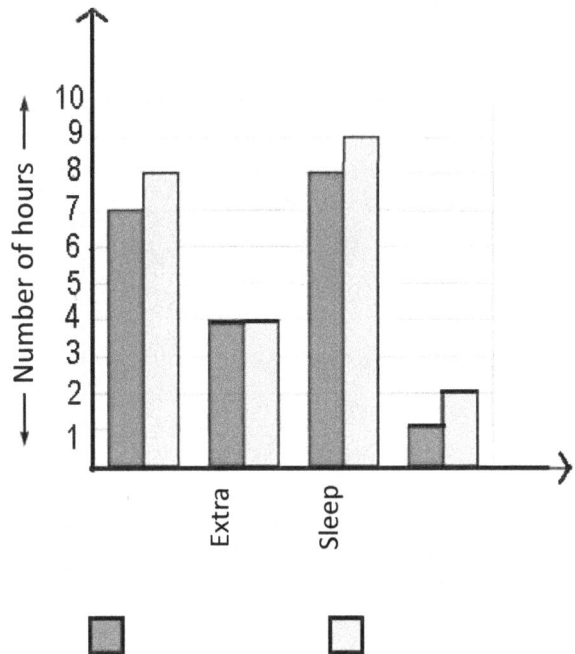

Fig:

Observation

Sleeping		

Result

Pie Chart

Aim

and represent it on a *pie chart*.

Materials Required

Prerequisite Knowledge

– The measure of the angle around a point is 360°.

Procedure

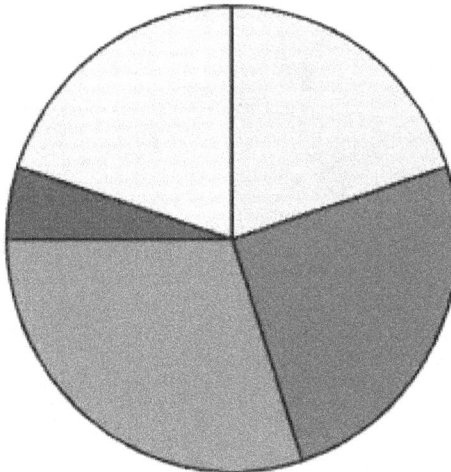

Fig:

Tennis

Observation

Name		$\dfrac{\text{Number of people}}{50} \times 360$
Tennis		

Result

Histogram

Aim

and represent it on a *histogram*.

Materials Required

Prerequisite Knowledge

–

rectangles of *equal widths close to each other*. The *height of the rectangles* corresponds to the *frequency* and the base corresponds to the *class* of the grouped data.

Procedure

possess.

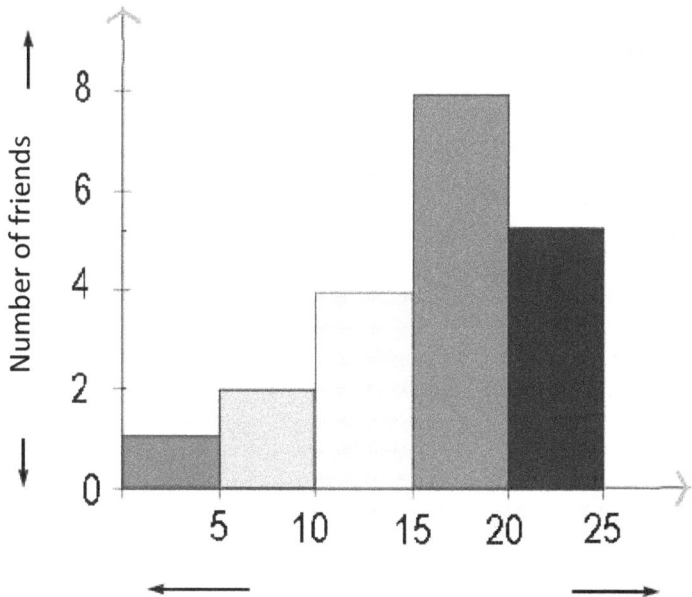

on the *x-axis* and the   on the *y-axis*.

friends possessing the maximum number of

Fig:

Observation

Result

Probability

Aim

To write the *sample space* and   of various

Materials Required

Prerequisite Knowledge

– The
sample space.

$$Probability\,of\,an\,event = \frac{Number\,of\,favourable\,outcome}{Total\,number\,of\,outcome}$$

Procedure

which defeats the

item).

scissor and paper.

Scissor

forward their hand to
create either a
paper or a *scissor*.
The winner of a

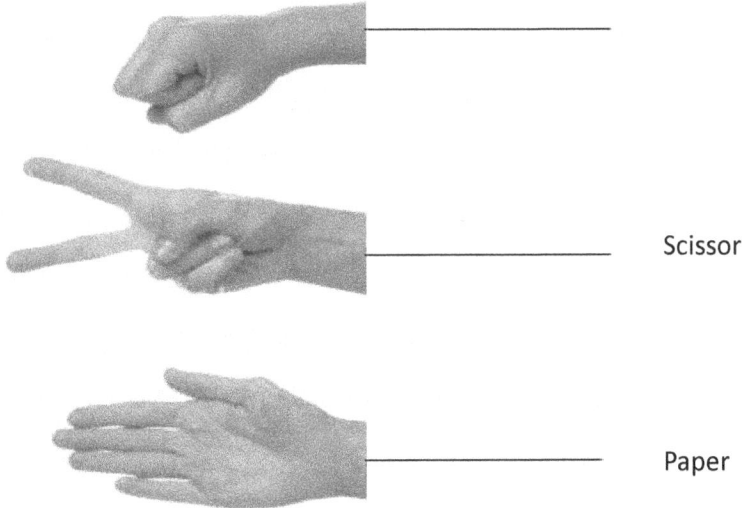

Paper

Fig:

Scissors defeats Paper.

Observation

		Winner/Tie

Sample space = _____

Result

Mean of Numbers

Aim

Materials Required

Prerequisite Knowledge

Procedure

(cubes) to show
numbers, 8,6,7,8 and 6.

number of cubes.

rearrangement denotes
the **mean**
it is seven.

formula of mean.

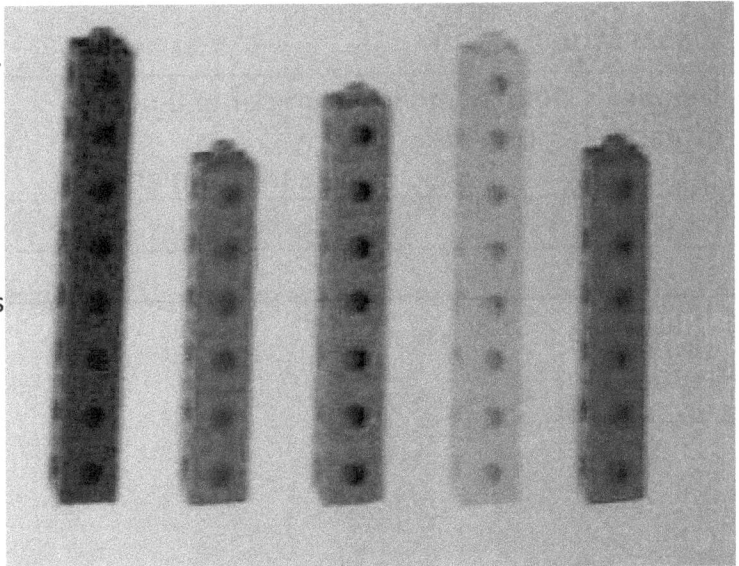

Fig:

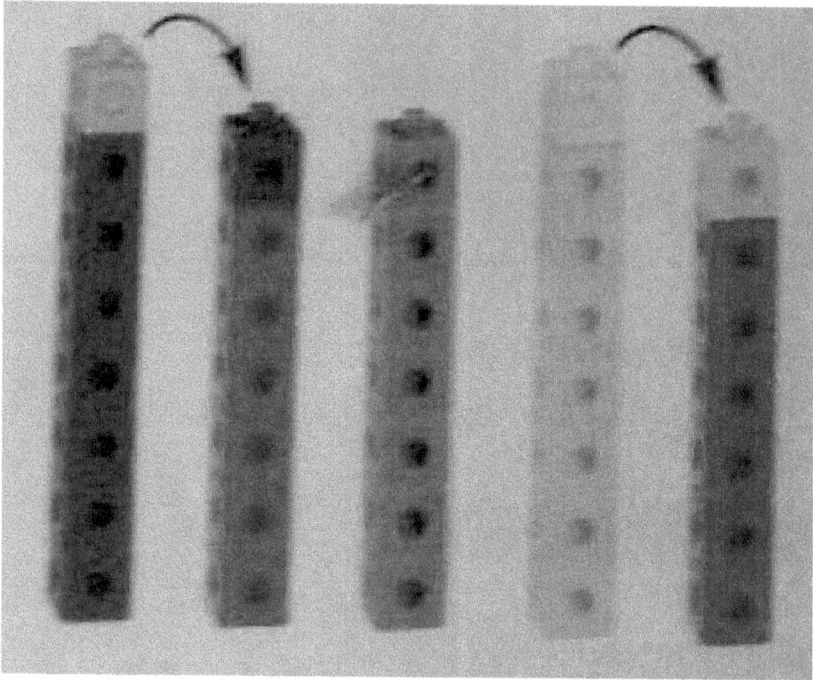

Fig:

Observation

Number of stacks	Total number of blocks	Number of blocks rearrangement	$\dfrac{\text{Total blocks}}{\text{Number of stacks}}$

Result

71 + 10 New Mathematics Projects

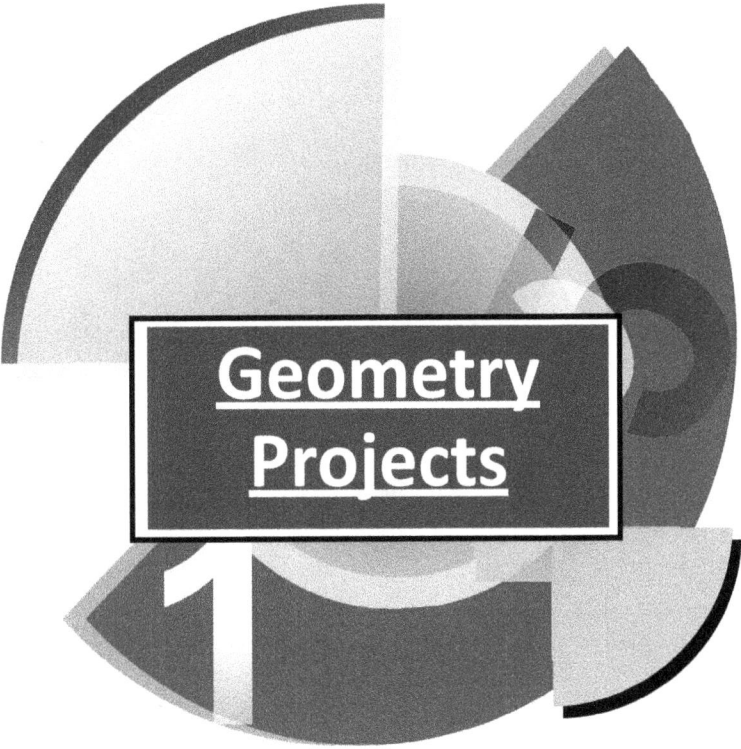

Geometry Projects

1

Tangram

Aim

Materials Required

Prerequisite Knowledge
–   made out of *a square* divided into   .
                                                                                                                                                   and
various *designs.*
Procedure

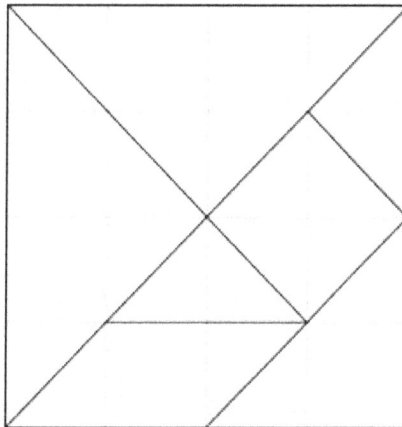

Fig: Tangram template

Observation

Name of the	Geometrical	Number of

Result

Line and Line Segment

Aim

To model line segments and lines using straws to exhibit an

Materials Required

Drawing sheet, ruler, pencil, straws (ten), glue, a pair of scissors.

Prerequisite Knowledge

has no

line segment is a part of a line that has two end points, i.e., it has a *point* and an end *point.*

— *misleading image*

Procedure

line segment.

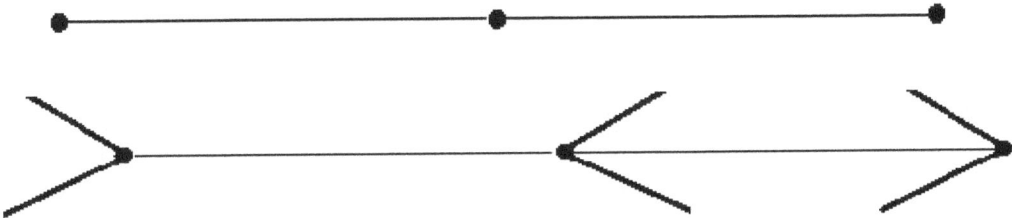

Fig: Line and line segment

Observation

Result

Angles formed by Parallel Lines

Aim

the   and *the alternate interior angles on the same side* of the transversal are *supplementary.*

Materials Required

Drawing sheet, ruler, pencil, thin cardboard, a pair of scissors, paper fastener.

Prerequisite Knowledge

—

—Two congruent angles which lie on the same side of the transversal are

alternate angles.

°.

Procedure

Draw two parallel lines, *'m'* and *'n'*   *'t'.*

fastener.

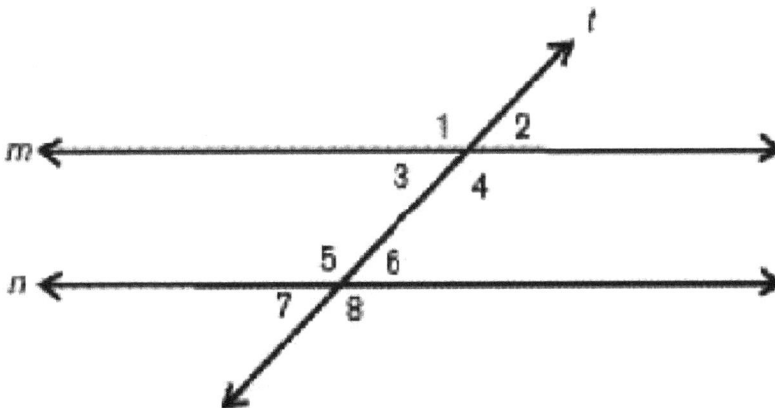

Fig:

Angles formed by Parallel Lines

Fig:

Observation

Angle number	Measure of the angle
1	
2	
3	
4	
5	
6	
7	
8	

Corresponding angles:

$\angle 1 =$ ——— $\angle 3 =$ ———

$\angle 2 =$ ——— $\angle 4 =$ ———

Vertically opposite angles:

$\angle 1 =$ ——— $\angle 5 =$ ———

$\angle 2 =$ ——— $\angle 6 =$ ———

Interior angles on the same side of the transversal:

$\angle 4 + \angle 6 =$ ———

$\angle 3 + \angle 5 =$ ———

Result

71 + 10 New Mathematics Projects

Angles on Magnetic Compass

Aim

Materials Required

Prerequisite Knowledge

    –   °and 90°   *acute angle.*

    –   °   *right angle.*

    –   °and 180°   .

    –   °   *straight angle.*

Procedure

the

Fig:

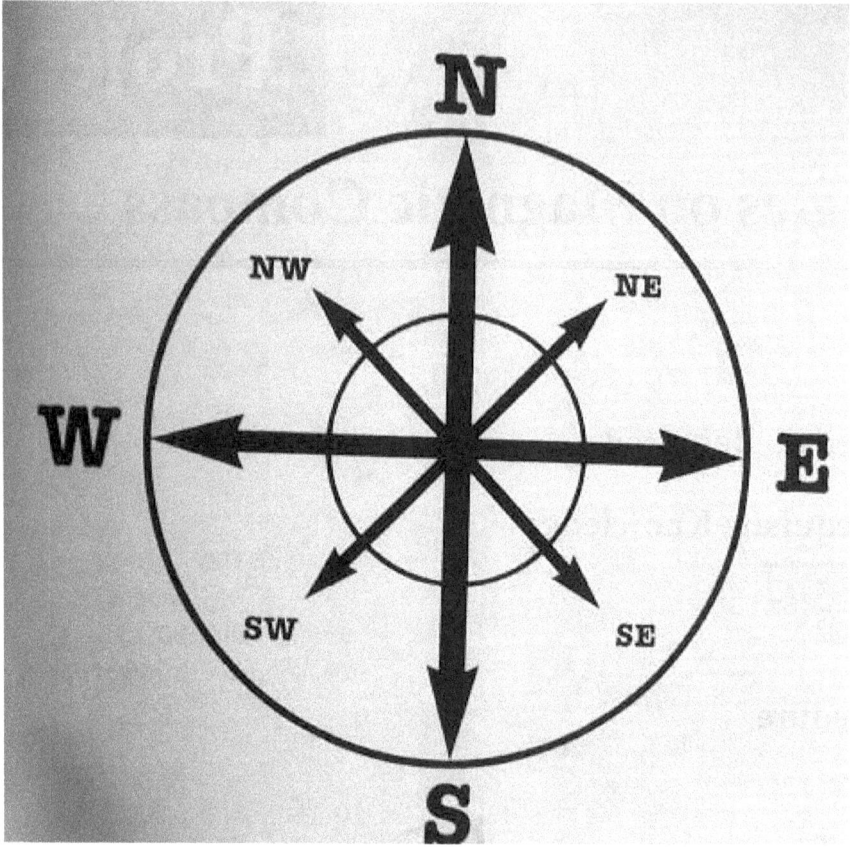

Fig:

Observation

Serial number	(between the arrows of)	
1.		
2.		
3.		
4.		
5.		
6.		

Result

Types of Triangles

Aim

equilateral and *isosceles triangle* can be and *acute* angled.

Materials Required

Ten straws, ten paper clips, a pair of scissors, ruler.

Prerequisite Knowledge

Isosceles triangle- An isosceles triangle is a triangle in which *two sides are equal.*

Equilateral triangle - An equilateral triangle is a triangle in which *three sides are equal.*

Right triangle - A right triangle is a triangle which has *one right angle.*

Obtuse triangle- An obtuse triangle is a triangle which has one *obtuse angle.*

Acute triangle- An acute triangle is a triangle in which all the angles are less than 90 degrees, so all the angles are *acute angles.*

Procedure

Now cut two straws of length 10 cm and one straw

isosceles triangle.

equilateral and *isosceles triangles.*

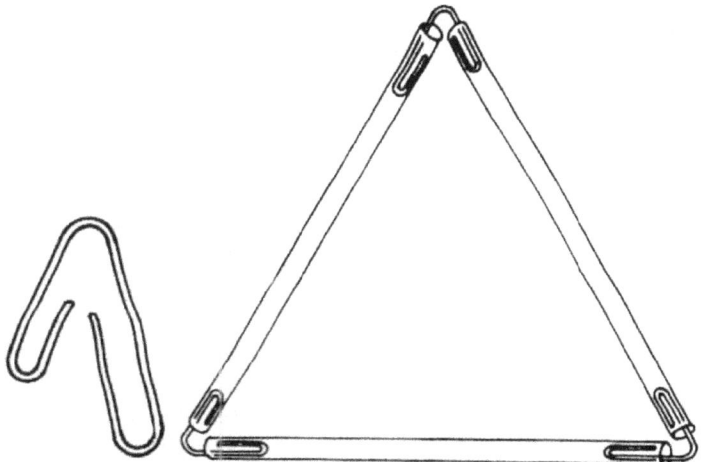

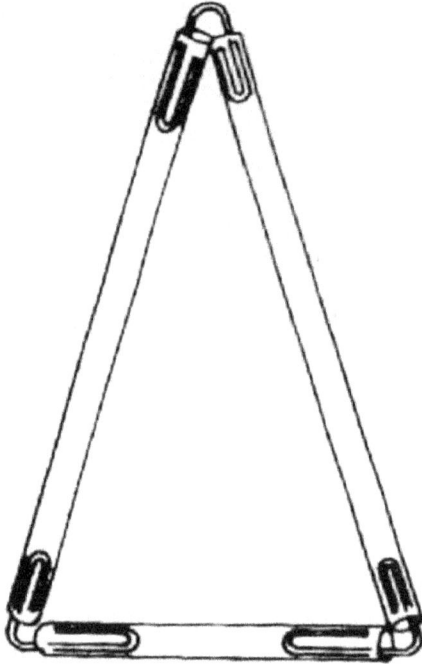

Fig: Straw triangles

Observation

Cases	based on side	angles	Name of the triangle based on angles	Name of the triangle based on both side and angle

Result

Angle Sum Property of a Triangle

Aim

Materials Required

Prerequisite Knowledge

–   °   *straight angle.*

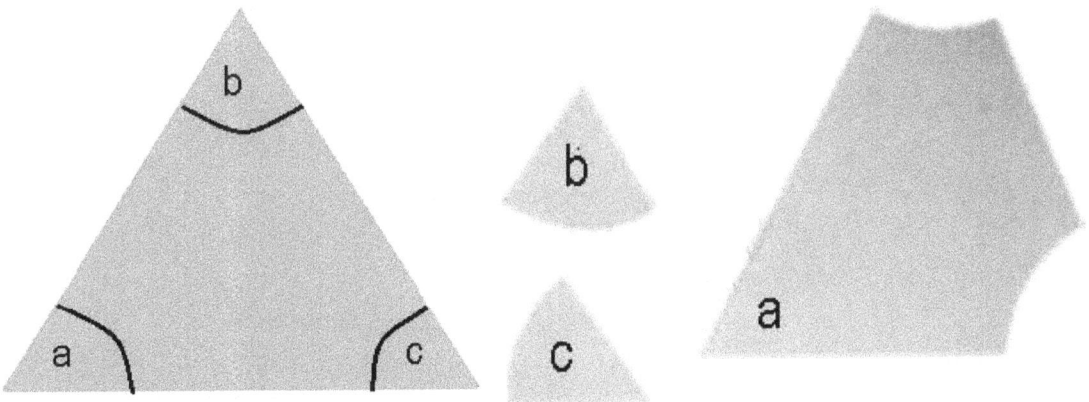

Procedure

them as '   and '*c*'.

and angle '*c*'.

Draw a straight line on the drawing sheet.

between the angles).

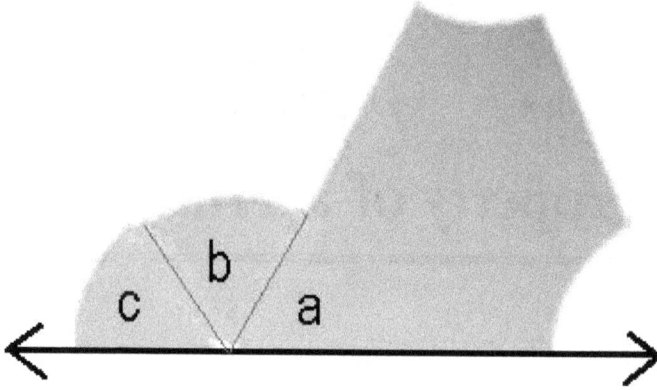

Fig:

Observation

Straight line = Straight angle = _____.

$\angle a + \angle b + \angle c =$ _____.

Result

Sum of Length Property of a Triangle

Aim

Materials Required

Prerequisite Knowledge

— *six*

elements of a triangle.

Procedure

sheet.

drawing sheet.

drawing sheet.

6 cm

8 cm

12 cm

14 cm

15 cm

Fig:

Observation

Length of straws		

Result

71 + 10 New Mathematics Projects

Median of a Triangle

Aim

Materials Required

Prerequisite Knowledge

–

median.

Procedure

equilateral triangle

median.

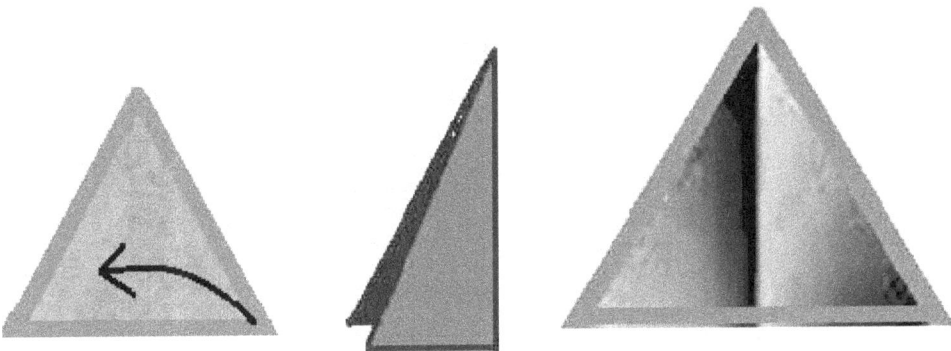

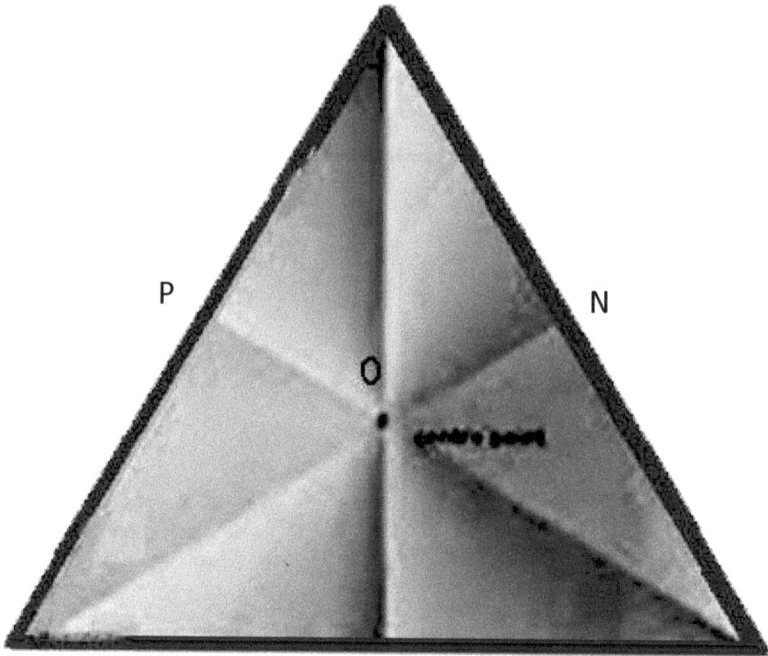

Fig:

Observation

Result

Project - 52

Area of Congruent Triangles

Aim

but if two triangles have the same area,

Materials Required

Prerequisite Knowledge

– Two triangles are said to be congruent if their corresponding sides and the

Procedure

another.

Now draw an acute angled triangle on the

Draw a right angled triangle of the same area.

congruent
another.

area of the triangles

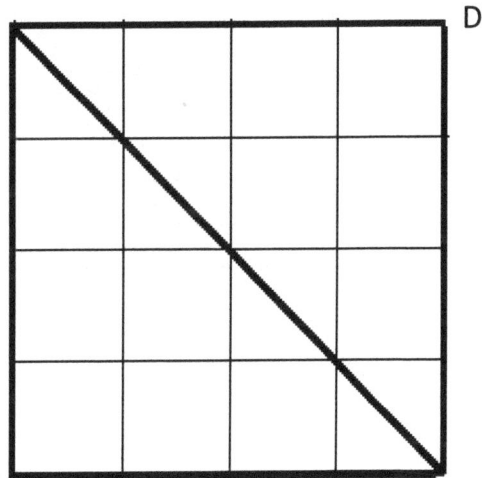

D

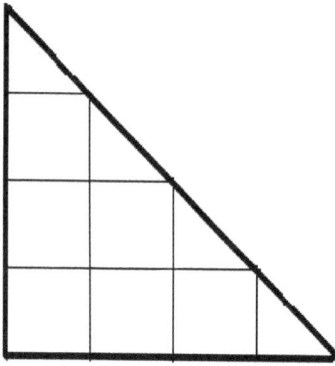

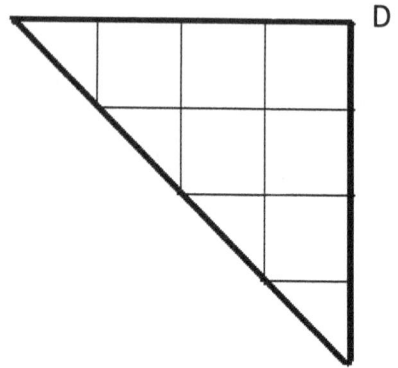

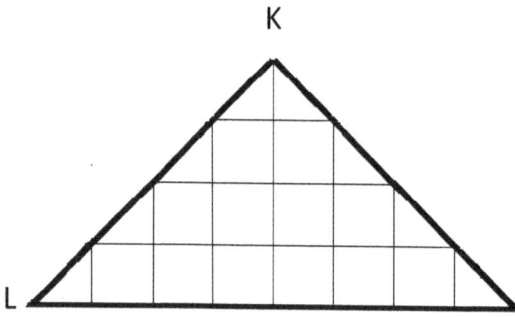

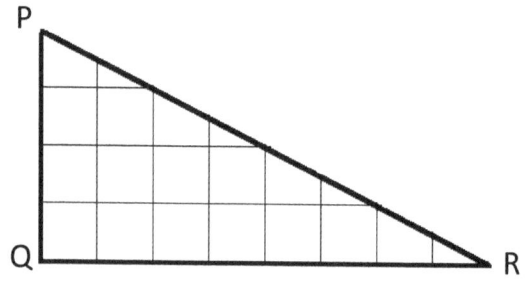

Fig:

Observation –

Name of the triangle	Number of unit squares

Result

71 + 10 New Mathematics Projects

Pythagoras Theorem

Aim

Pythagoras Theorem   *unit squares.*

Materials Required

Prerequisite Knowledge

–

Procedure

other two sides as 4 units and 3 units.

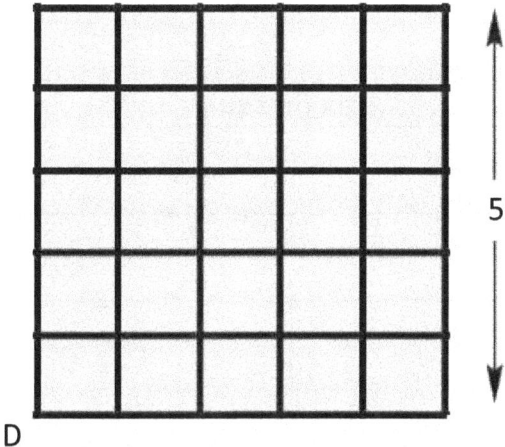

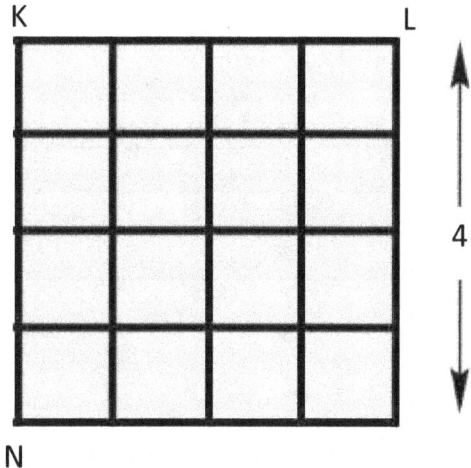

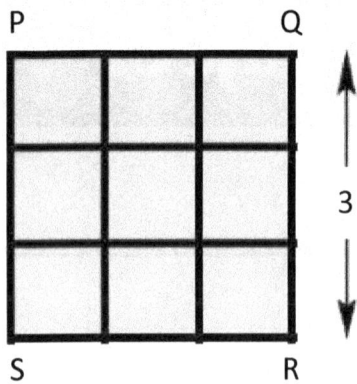

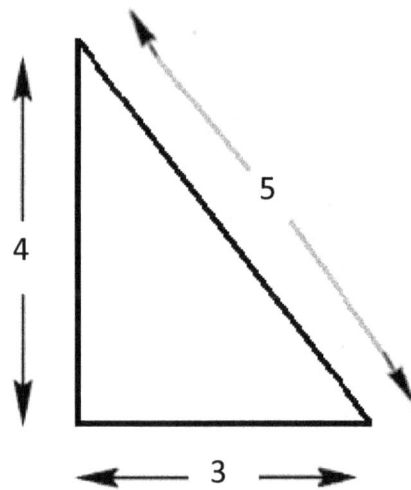

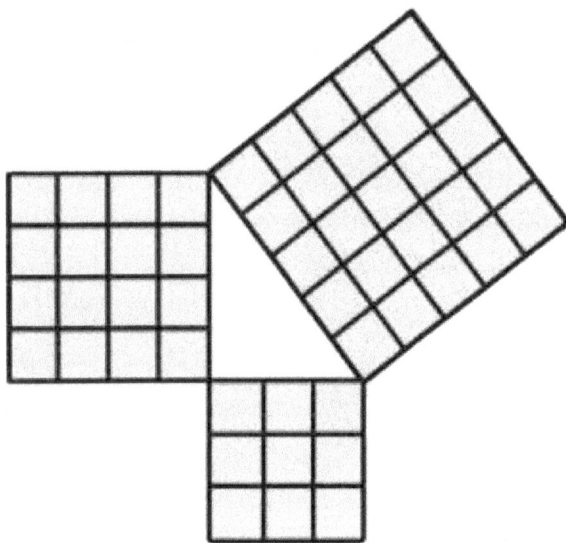

Fig:

Observation

Name of the square	Area (in square units)

Result

Hexagon from a Circle

Aim
To draw a *hexagon* inside a circle and cut it into pieces to form a *rectangle*.

Materials Required

Prerequisite Knowledge

–

        radius.

   –   *perimeter* of the circle is called its *circumference.*

Procedure

Draw a circle of radius eight

Keep the compass point on the

an arc on the circumference (do not change the radius).

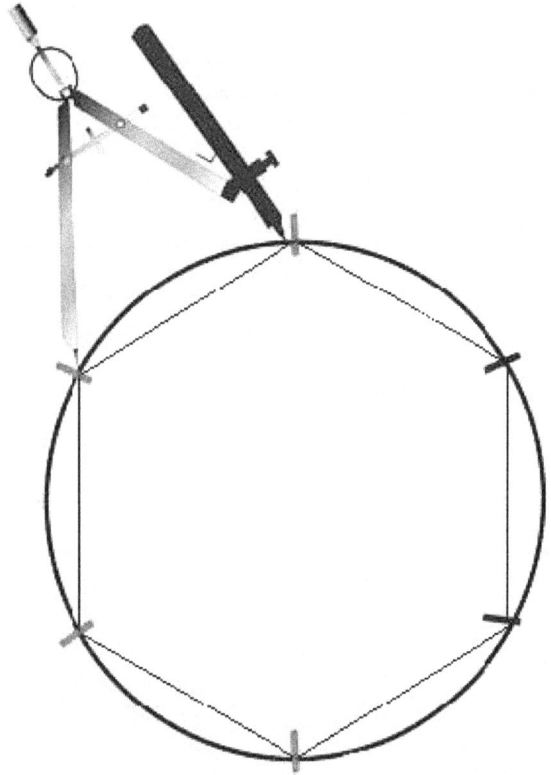

changing the radius.

made on the circumference.

parts and rearrange them to form a *rectangle*.

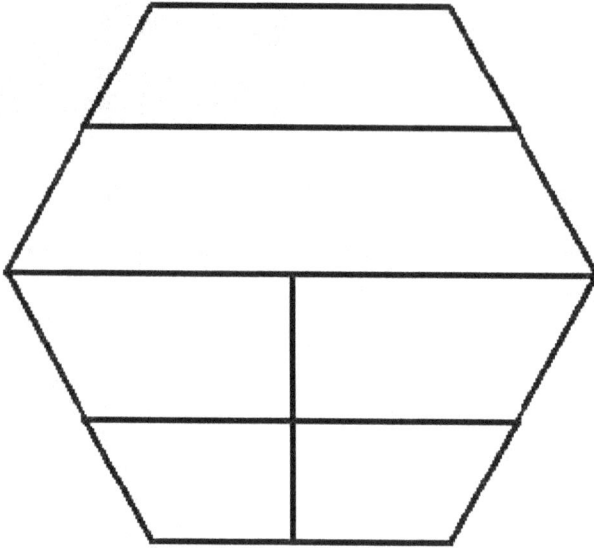

Fig: Hexagon from a circle

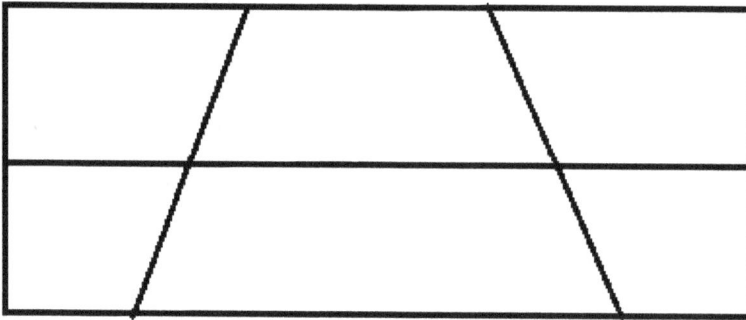

Fig: Rearrangement of hexagon pieces

Observation

Result

Angle Sum Property of a Quadrilateral

Aim

Materials Required

Prerequisite Knowledge

– $^{\circ}$ *an angle around a point.* This

is also called a *complete angle* since this completes

Procedure

them as '   '*c*' and '*d*'.

'*a*', angle '*c*' and angle '*d*'.

(do not leave space between the angles).

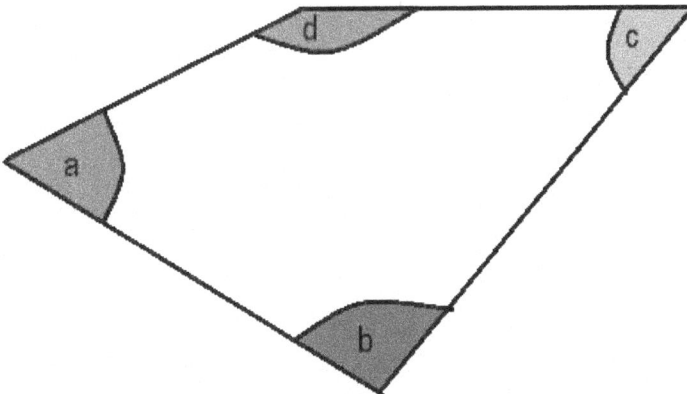

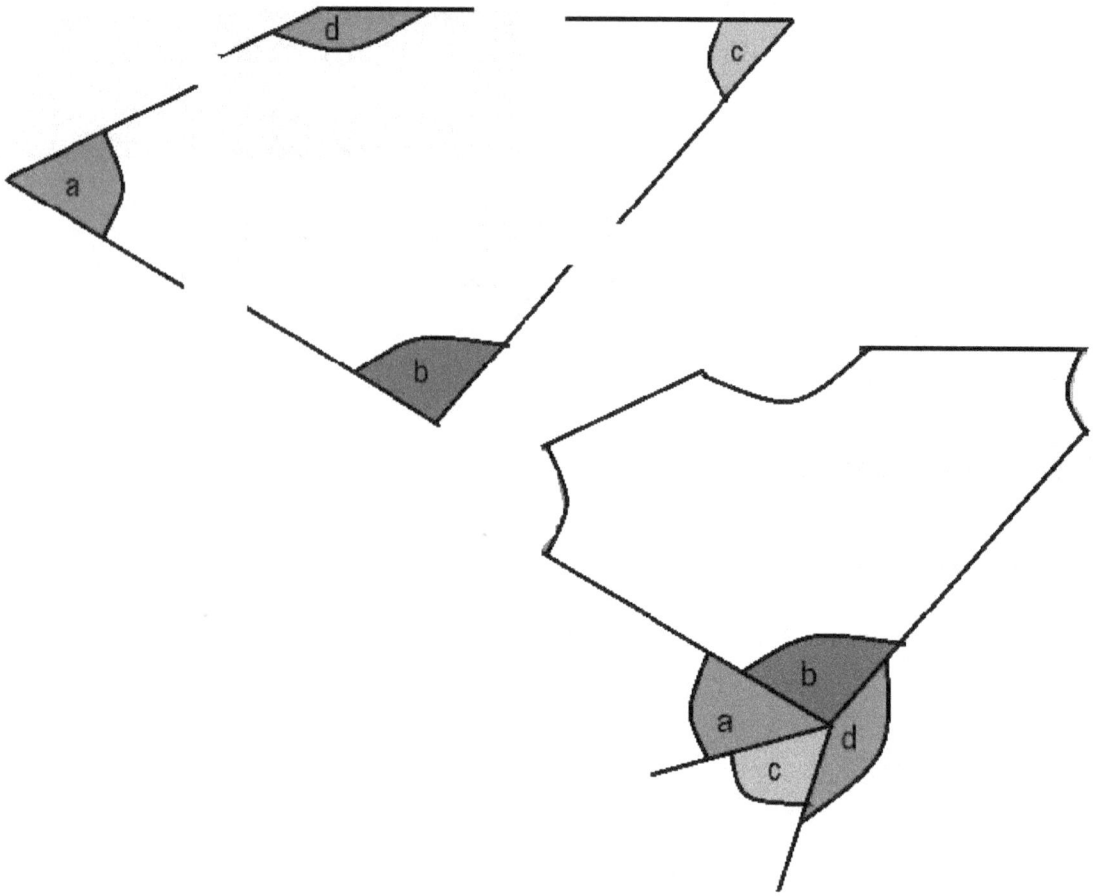

Fig:

Observation

$$\angle a + \angle b + \angle c + \angle d = \underline{\hspace{3cm}}$$

Result

Area and Perimeter of a Square

Aim
To derive the formula of *area and perimeter*

Materials Required

Prerequisite Knowledge
–

perimeter. Perimeter

Area –

area.

Procedure

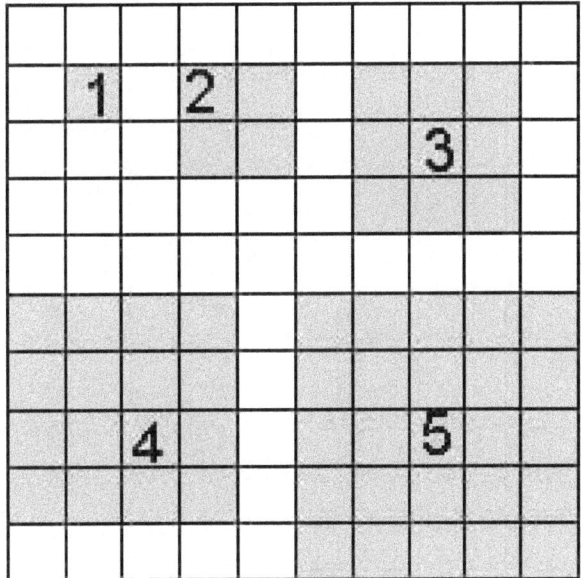

Fig:

Observation

Figure	Length (in units)	Number of unit squares	Area (in square units)	Sum of sides (in units)	Perimeter (in units)

Result

71 + 10 New Mathematics Projects

Area and Perimeter of a Rectangle

Aim
To derive the formula of the *area* and *perimeter* of a rectangle.

Materials Required

Prerequisite Knowledge
–

perimeter. The Perimeter

Area –

Area.

Procedure

the form of rectangles as

of each of the four rectangles.

table.

between the length and breadth of the rectangles and their

between the length and breadth of the rectangles and their perimeter.

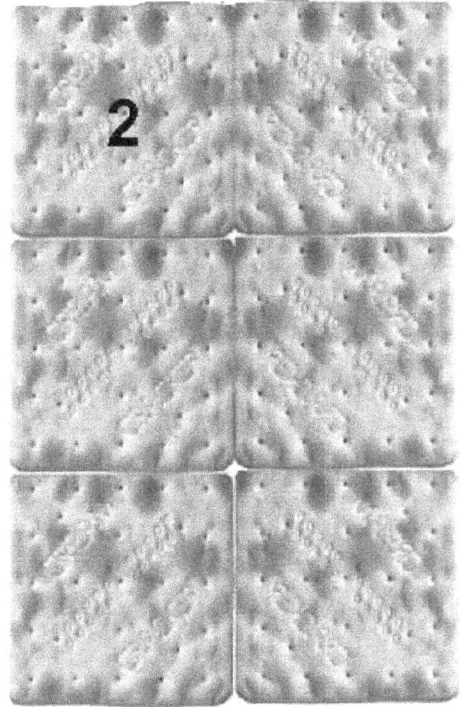

Length

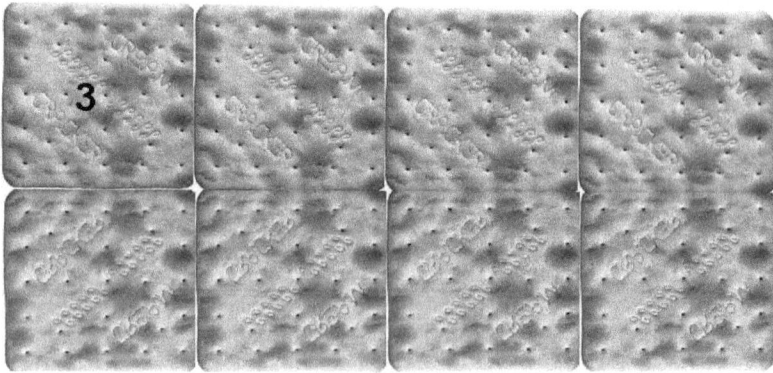

Fig:

Observation

Figure	Length (in units)	Breadth (in units)	Number of unit squares	Area (in square units)	Sum of four sides (in units)	Perimeter (in units)

Result

Area of a Parallelogram

Aim
To derive the formula of the area and perimeter of a parallelogram.

Materials Required

Prerequisite Knowledge
—

parallelogram.

Procedure

thin cardboard and cover it with a

Draw a perpendicular from one of

This represents the height of the parallelogram.

a *right angled triangle* and a *quadrilateral.*

Place the triangle on the opposite

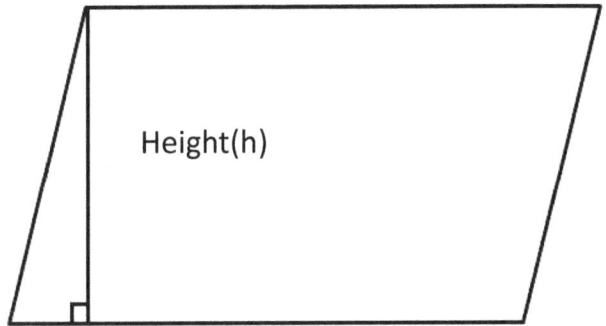

Height(h)

the area of the new

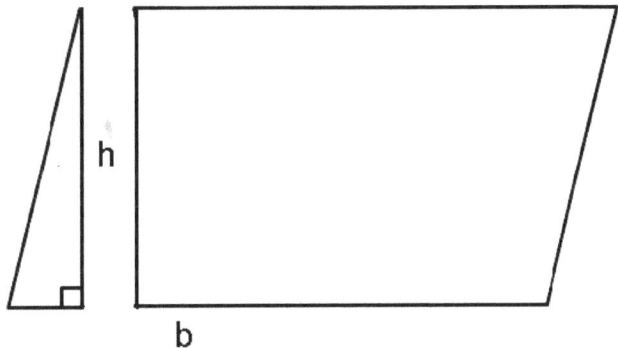

h

b

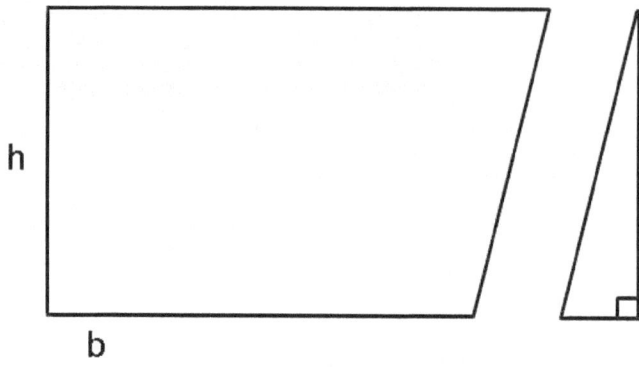

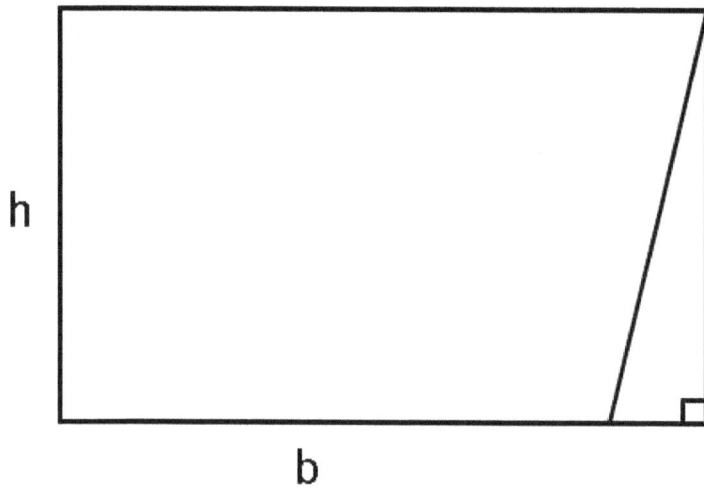

Fig:

Observation

Result

Area of a Triangle

Aim

To derive the formula of the area of a triangle.

Materials Required

Prerequisite Knowledge

Procedure

paper.

represents the height of the triangle.

∘.

of a triangle.

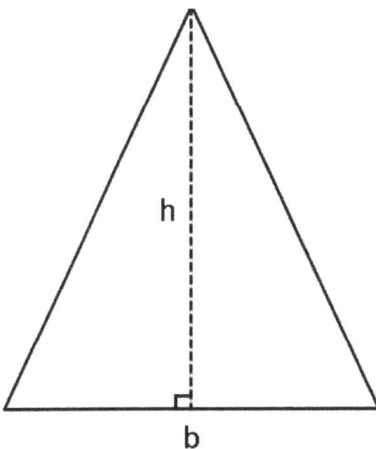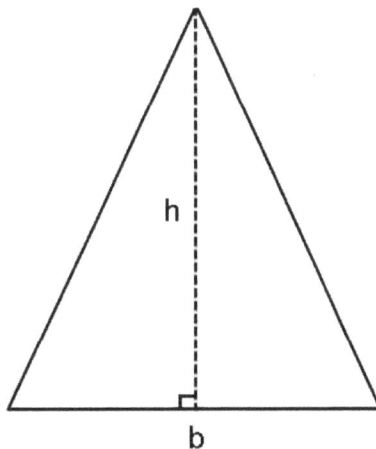

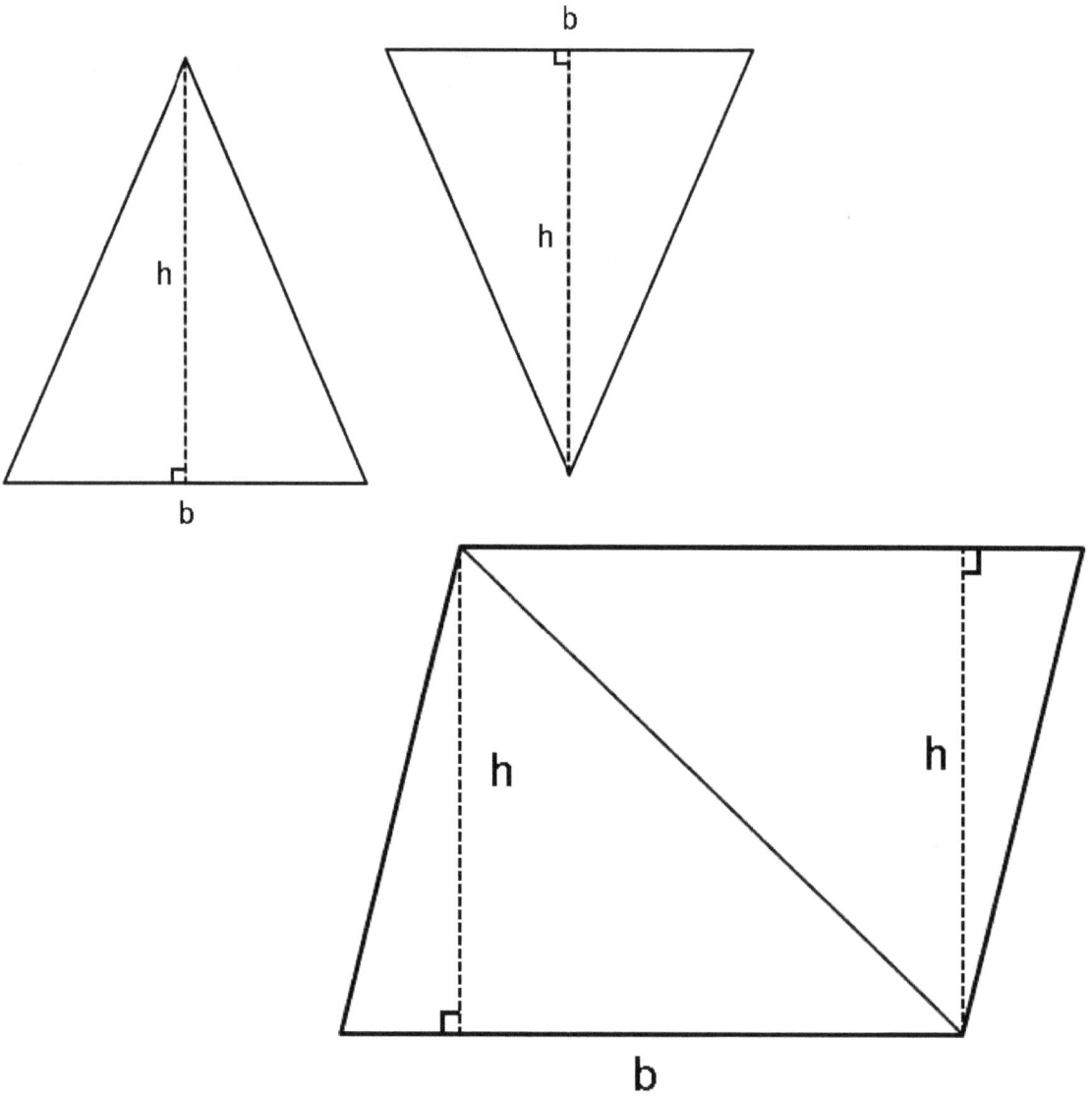

Fig:

Observation

Result

71 + 10 New Mathematics Projects

Same Area Different Perimeters

Aim

To calculate the *area* and *perimeter* of a *square* and *rectangle* from a *square grid*.

Materials Required

Prerequisite Knowledge

Perimeter of any shape = Sum of all the sides

Procedure

adding all the sides.

perimeters.

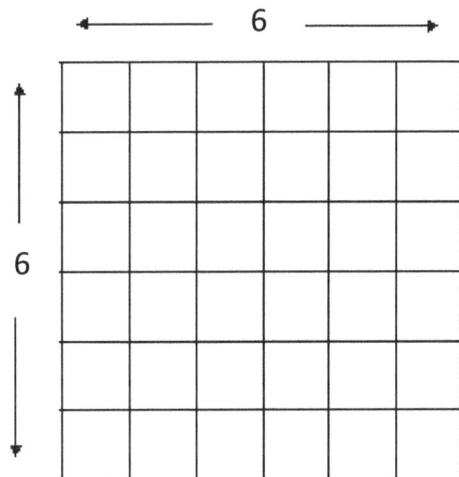

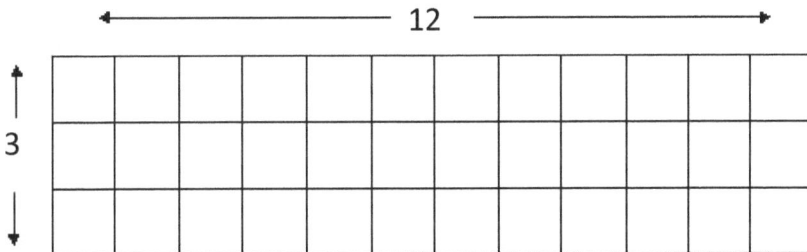

Fig:

Observation

	Area(in cm²)	Perimeter (in cm)

Result

Properties of Square and Rhombus

Aim

Materials Required

Prerequisite Knowledge

—

–

Procedure

square
Draw two diagonals, PR and QS.

thread where it meets Q.

thread to measure all the four sides.

diagonals.

earlier.

Repeat the same steps for a rhombus.

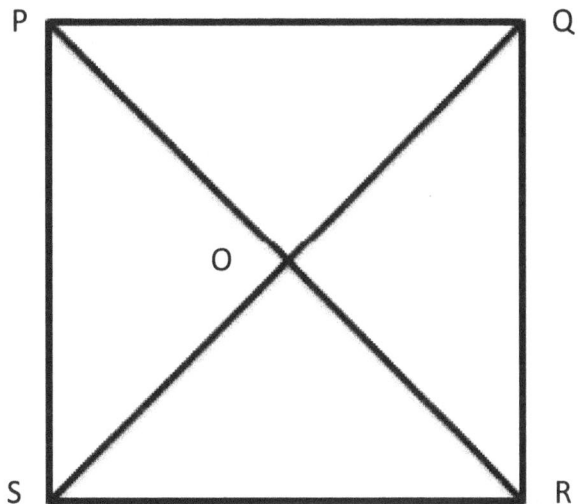

P ▭ Q

O

S ▭ R

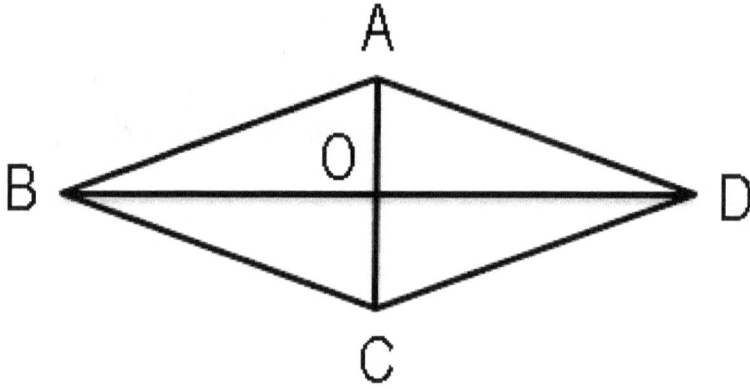

Fig:

Observation

	Square	Rhombus
Length of four sides		
Parallel sides		
Length of diagonals		

Result

71 + 10 New Mathematics Projects

Properties of Parallelogram and Rectangle

Aim

parallelogram and *a rectangle.*

Materials Required

Prerequisite Knowledge

—

-

Procedure

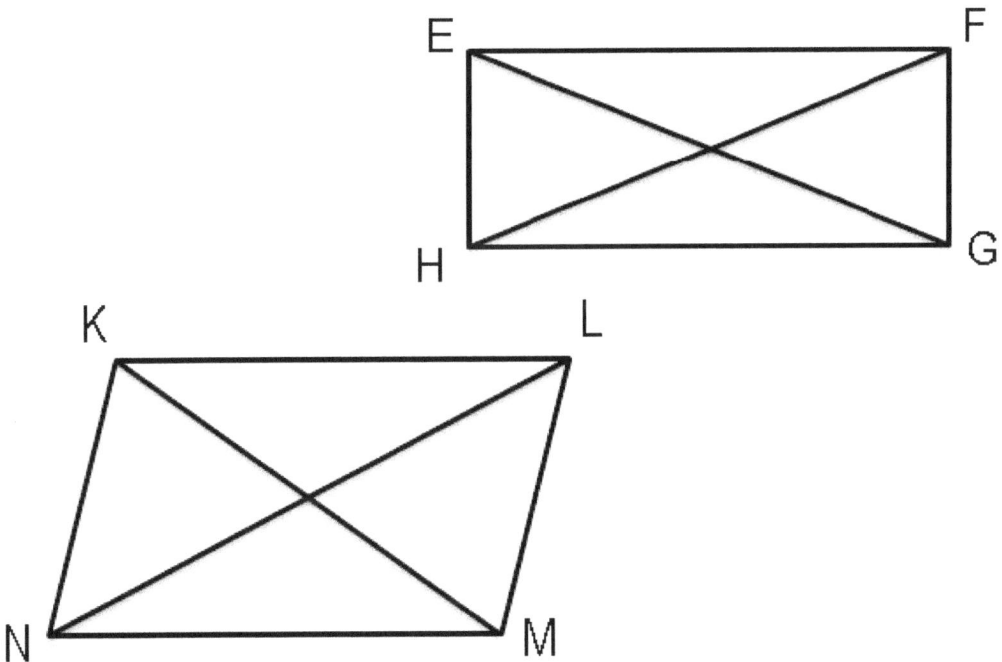

Fig:

Observation

	Rectangle	Parallelogram
Length of opposite sides		
Parallel sides		
Length of diagonals		

Result

Pi is a Constant

Aim

To show that   has a   .

Materials Required

dial.

Prerequisite Knowledge

Diameter –
as a *diameter.*

–   *circumference.*

Procedure

Diameter

circular plate, etc.

circumference and
the diameter of each
object.

of pi   *each*
circumference
the *corresponding*
diameter.

Fig:

Observation

Name of the object	Circumference	Diameter	Circumference / Diameter

Result

Area of a Circle

Aim

formula of the *area* of a *circle*.

Materials Required

Drawing sheet, compass, pencil, a pair of scissors

Prerequisite Knowledge –

- *two radii* and an *arc* *sector*.

Procedure –

Draw a circle of radius 6 cm.

form a parallelogram).

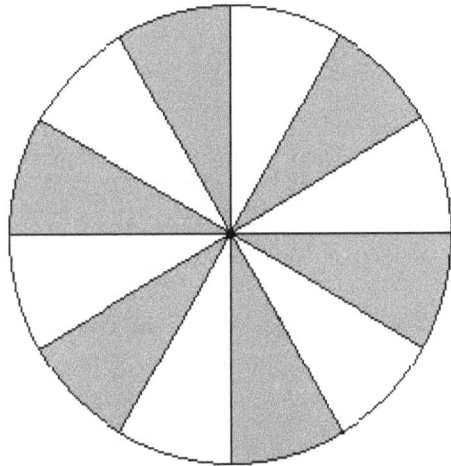

parallelogram.

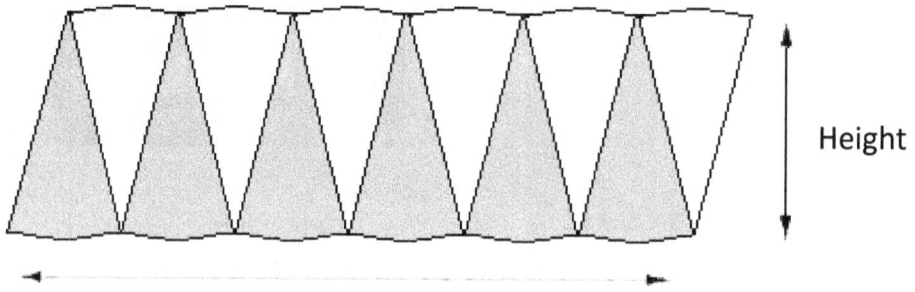

Height

Fig: Sectors of a circle and rearrangement of sectors into a parallelogram

Observation

_____ of the parallelogram = _____ of the circle = _____ .

_____ of the parallelogram = _____ × _____ of the circle.

Result

Surface Area of a Cube

Aim

and the *total surface area* of a cube.

Materials Required

Thin cardboard, a pair of scissors, glue, tracing or carbon paper, pencil.

Prerequisite Knowledge

—

—

the

Procedure

Trace or draw the net of the cube on a thin cardboard.

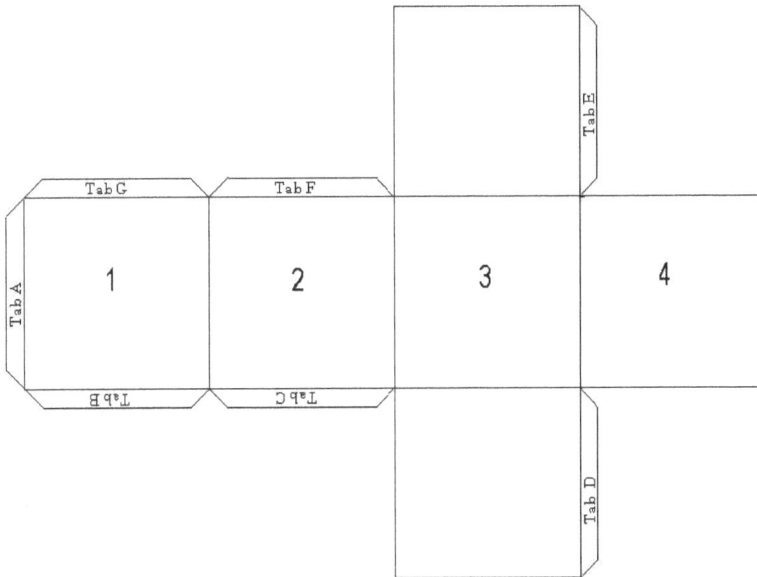

Fig: Net of a cube

Observation

Face	Area (in square units)
1	
2	
3	
4	
Top	

= _____

= _____

Result

Animal Shapes using Unit Cubes

Aim

animal shape using

Materials Required

Prerequisite Knowledge

Volume
in cubic units.

Procedure –

Fig: Dinosaur from unit cubes

Observation

Result

Surface Area Comparison

Aim

Materials Required
Thin cardboard, a pair of scissors, glue, tracing or carbon paper, pencil.

Prerequisite Knowledge
– The total area of all the faces along with the curved surface of a solid three

Procedure
Trace the net of the cube and draw it on a thin

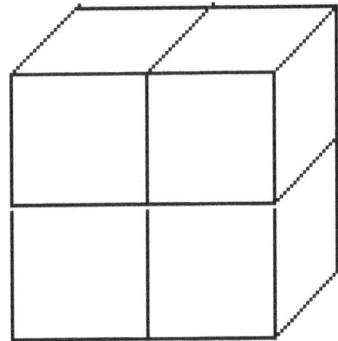

and glue the tabs to form a cube.

four such

Put the four cubes next to each other in a straight line and count the total number of the faces of the cubes.

other two cubes and again count the total number of the faces of the cubes.

the result.

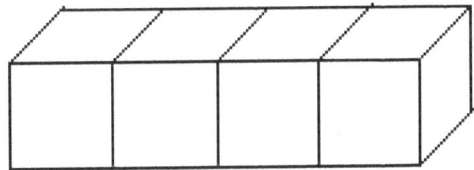

Fig:

Observation

the four cubes	Number of faces
Next to each other	
Two on top of the other two	

Result

Volume of Cube and Cuboid

Aim

formula of the   of a   and a

Materials Required

Prerequisite Knowledge

–

in

Procedure

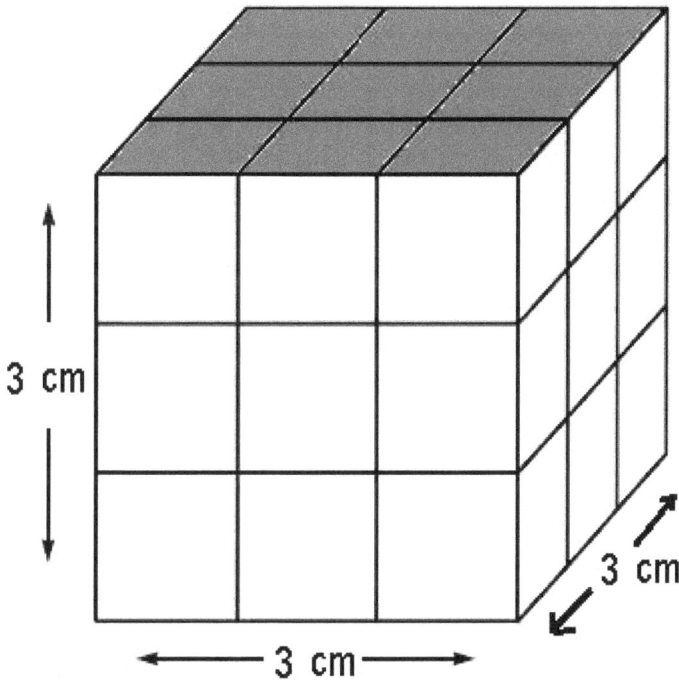

3 cm

3 cm

◄——— 3 cm ———►

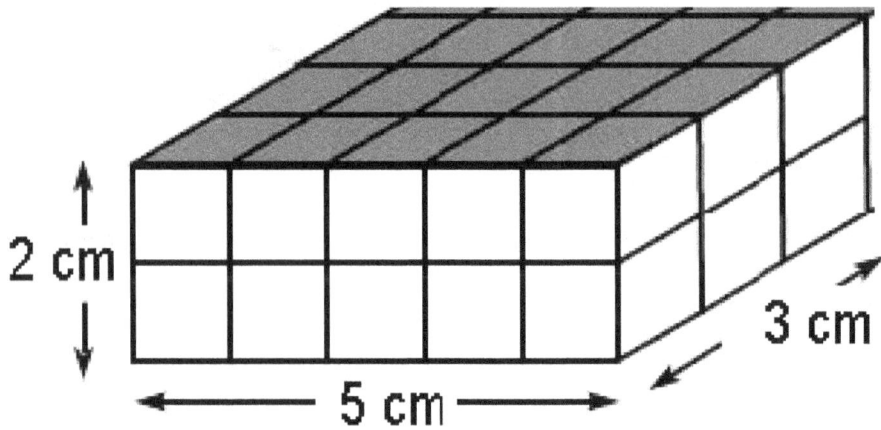

Fig:

Observation

Cube	Total number of unit cubes	Number of unit cubes on side$_1$	Number of unit cubes on side$_2$	Number of unit cubes on side$_3$	Side×Side×Side

Cuboid	Total number of unit cubes	Number of unit cubes on length	Number of unit cubes on breadth	Number of unit cubes on height	L × B × H

Result

71 + 10 New Mathematics Projects

Area of a Cylinder

Aim

and the *total surface area*

Materials Required

Drawing sheet, compass, pencil, a pair of scissors

Prerequisite Knowledge

2

Procedure

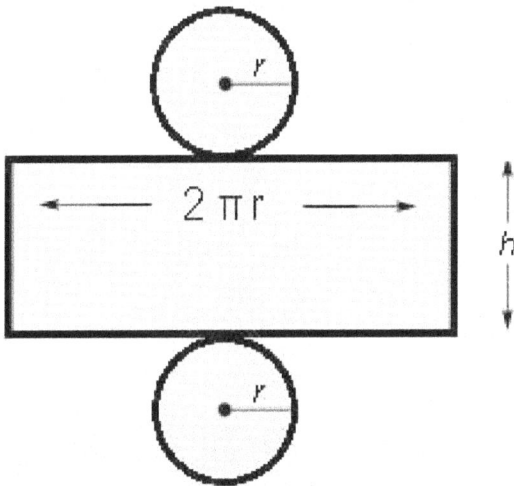

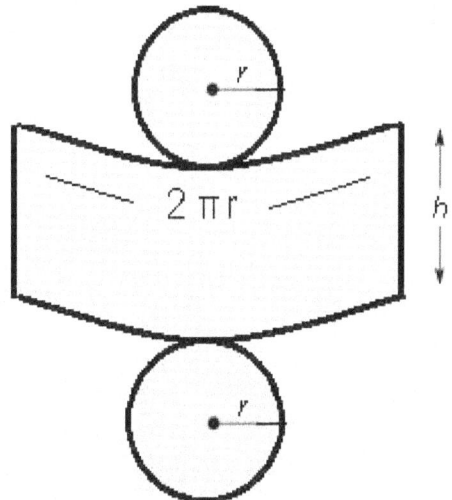

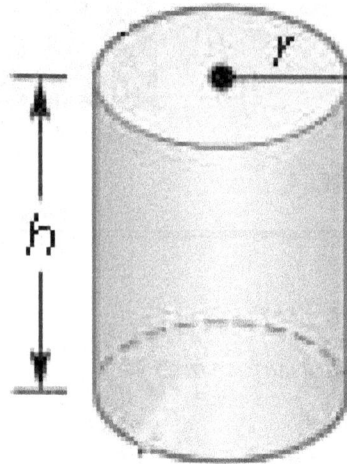

Fig:

Observation

= _____ × _____

= _____ + _____ + _____ = _____

Result

71 + 10 New Mathematics Projects

3D Shapes from the Net

Aim

edges and of a and a *square pyramid* and to *Euler's formula.*

Materials Required

Thin cardboard, a pair of scissors, glue, tracing or carbon paper, pencil.

Prerequisite Knowledge

— *three-dimensional geometric* with a *square* and *four triangular sides* that connect at one point is called a *square pyramid.*

Procedure

Trace the *net* of the *cylinder, cone* and *pyramid* and draw them on a thin cardboard.

the boundaries.

lines.

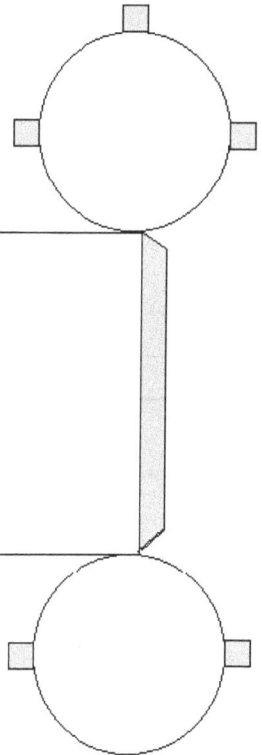

of *faces, edges* and . Note down

Fig:

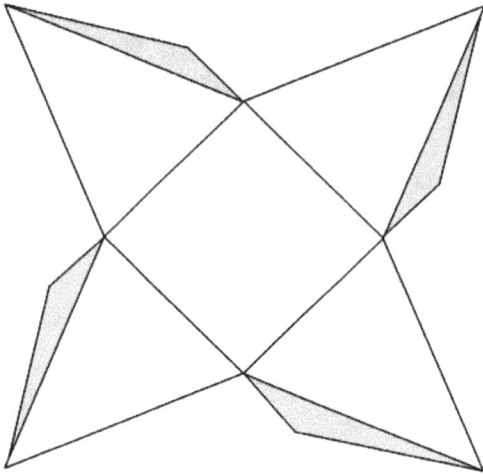

Fig: Net of a cone

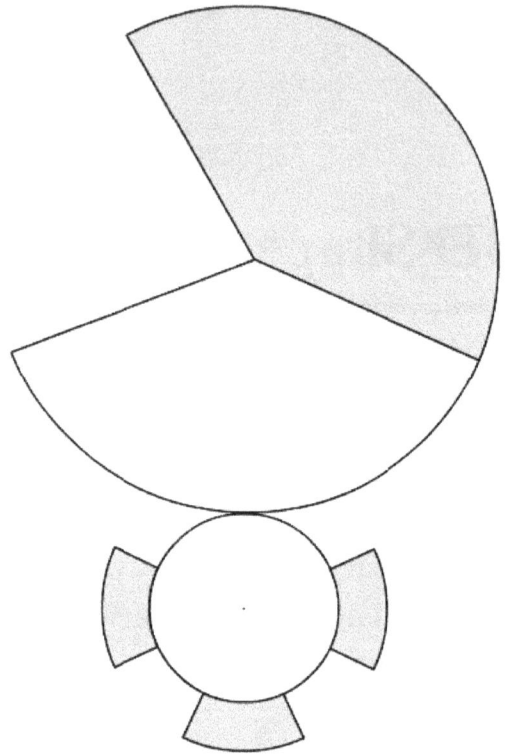

Fig:

Observation

	Number of faces	Number of edges	Number of	Euler's Formula: Number of faces + Number of

Result

71 + 10 New Mathematics Projects

Symmetrical Design

Aim

Materials Required

Prerequisite Knowledge

–   *congruent parts*   *line of*
symmetry.
–   *centre of the circle*   *circumference* of
                    radius.

Procedure

diameter.

central radius.

along the central radius.

from the pointed end and
four triangles from each of
the three sides of the folded

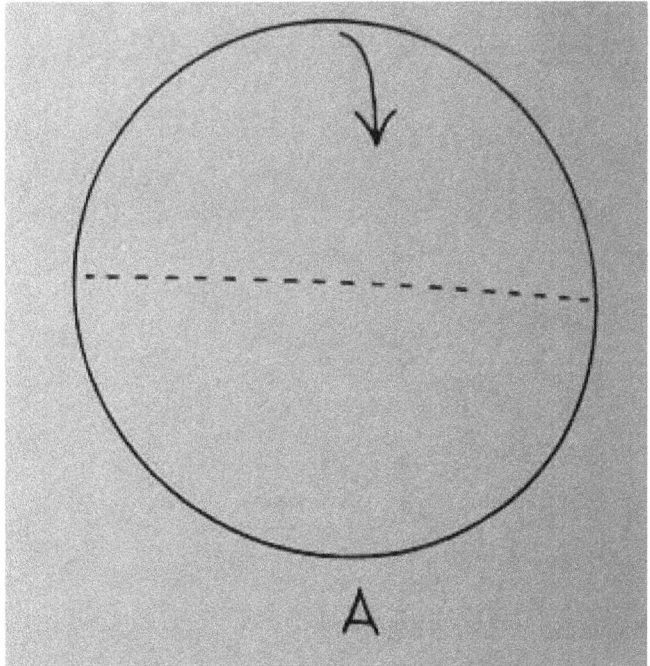

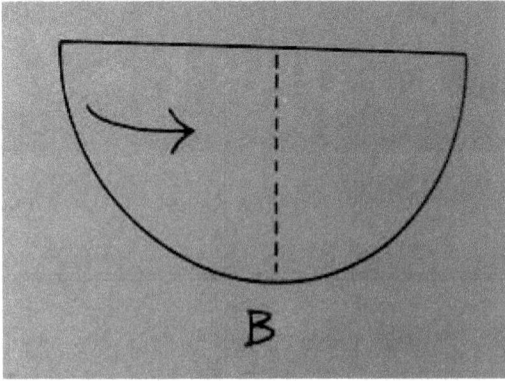

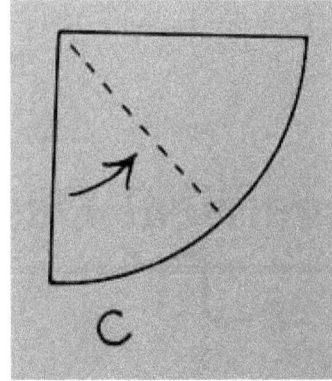

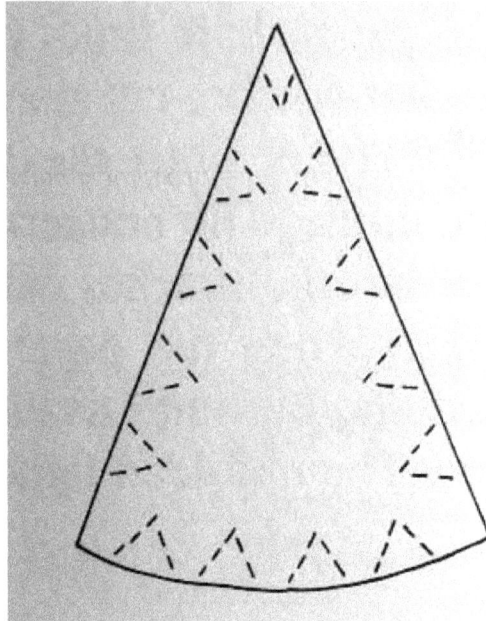

Fig:

Observation

Result

71 + 10 New Mathematics Projects

10 New Projects

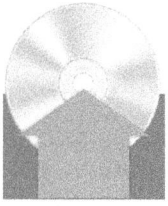

Square Numbers

Aim

To display square numbers on a square grid paper.

Materials required

Ruler, pencil, a pair of scissors, drawing sheet, square grid.

Prerequisite Knowledge

Square number – The product of some integer with itself is known as a *square number* or a *perfect number*.

Procedure –

Cut out one small square from the square grid. This gives the

Now shade three squares around one square in order to form a big square. This gives the *second square number.*

big square. This gives the *third square number.*

number at each step.

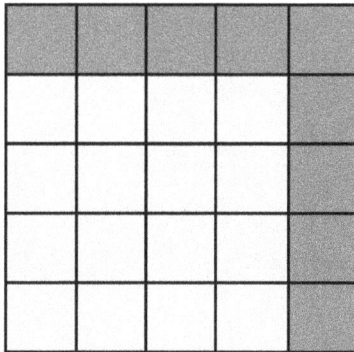

Fig. 25 = 1 + 3 + 5 + 7 + 9

Observation

Shape number	Square grid	Total number of unit squares	Total squares in terms of odd numbers/products

Result

Extension – Triangular numbers can be displayed in a similar manner.

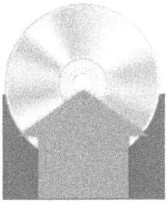

Equivalent Fractions

Aim

Materials required

Ruler, pencil, thin cardboard, a pair of scissors, glazed paper, sketch pen.

Prerequisite Knowledge

–

the number line.

Procedure

Cut out 12 rectangles of same length and breadth from a thin cardboard. Cover them with a glazed paper.

as it is. Divide the second rectangle into two equal parts, the third into three equal parts, the fourth into four equal parts and so on,

equal parts).

Cut out the equal parts which

to make a   and

of 1/4, 1/3, 1/2, 1/5, 3/4, 2/5 and 1/6.

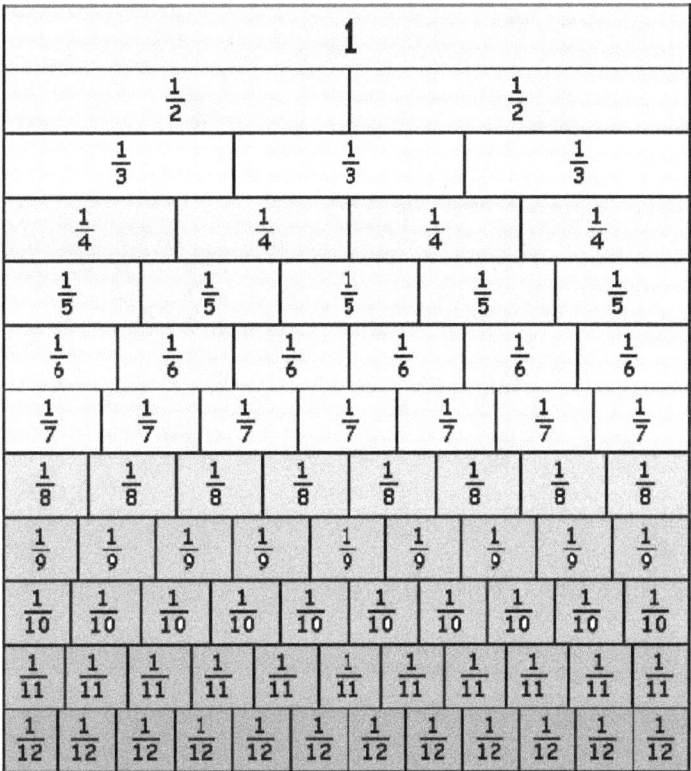

1

	$\frac{1}{2}$			$\frac{1}{2}$	

| | $\frac{1}{3}$ | | $\frac{1}{3}$ | | $\frac{1}{3}$ |

| $\frac{1}{4}$ | | $\frac{1}{4}$ | | $\frac{1}{4}$ | | $\frac{1}{4}$ |

| $\frac{1}{5}$ | $\frac{1}{5}$ | $\frac{1}{5}$ | $\frac{1}{5}$ | $\frac{1}{5}$ |

| $\frac{1}{6}$ | $\frac{1}{6}$ | $\frac{1}{6}$ | $\frac{1}{6}$ | $\frac{1}{6}$ | $\frac{1}{6}$ |

| $\frac{1}{7}$ | $\frac{1}{7}$ | $\frac{1}{7}$ | $\frac{1}{7}$ | $\frac{1}{7}$ | $\frac{1}{7}$ | $\frac{1}{7}$ |

| $\frac{1}{8}$ | $\frac{1}{8}$ | $\frac{1}{8}$ | $\frac{1}{8}$ | $\frac{1}{8}$ | $\frac{1}{8}$ | $\frac{1}{8}$ | $\frac{1}{8}$ |

| $\frac{1}{9}$ | $\frac{1}{9}$ | $\frac{1}{9}$ | $\frac{1}{9}$ | $\frac{1}{9}$ | $\frac{1}{9}$ | $\frac{1}{9}$ | $\frac{1}{9}$ | $\frac{1}{9}$ |

| $\frac{1}{10}$ | $\frac{1}{10}$ | $\frac{1}{10}$ | $\frac{1}{10}$ | $\frac{1}{10}$ | $\frac{1}{10}$ | $\frac{1}{10}$ | $\frac{1}{10}$ | $\frac{1}{10}$ | $\frac{1}{10}$ |

| $\frac{1}{11}$ | $\frac{1}{11}$ | $\frac{1}{11}$ | $\frac{1}{11}$ | $\frac{1}{11}$ | $\frac{1}{11}$ | $\frac{1}{11}$ | $\frac{1}{11}$ | $\frac{1}{11}$ | $\frac{1}{11}$ | $\frac{1}{11}$ |

| $\frac{1}{12}$ | $\frac{1}{12}$ | $\frac{1}{12}$ | $\frac{1}{12}$ | $\frac{1}{12}$ | $\frac{1}{12}$ | $\frac{1}{12}$ | $\frac{1}{12}$ | $\frac{1}{12}$ | $\frac{1}{12}$ | $\frac{1}{12}$ | $\frac{1}{12}$ |

Fig.

Equivalent Fractions

Observation

Result

Multiplication of Decimals

Aim

square grid.

Materials required

Ruler, pencil, drawing sheet, crayons, square grid.

Prerequisite Knowledge

Procedure

The squares coloured with both the colours represent the product.

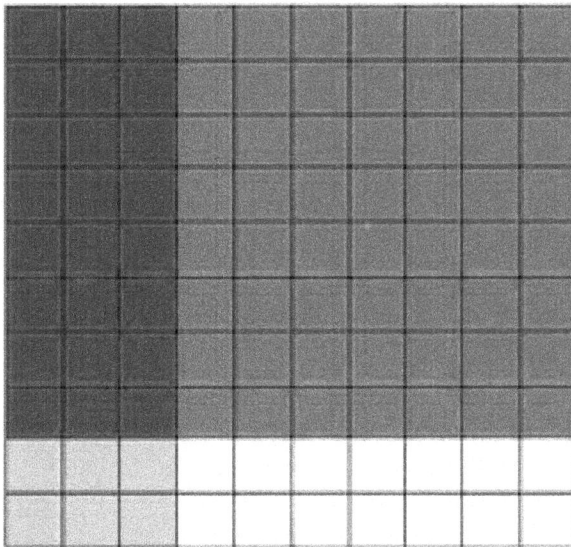

Fig.

Observation

Problem	First factor square grid	Second factor square grid	Product square grid	Product

Result

Scale Drawing

Aim

To draw the _____ on a *square grid*.

Materials required

Square grid, ruler, sketch pen.

Prerequisite Knowledge

Scale Drawing – An enlarged or reduced drawing of an actual object is known as _____ .

Procedure

Determine the _____ For example, 1 unit square = 1.5 m

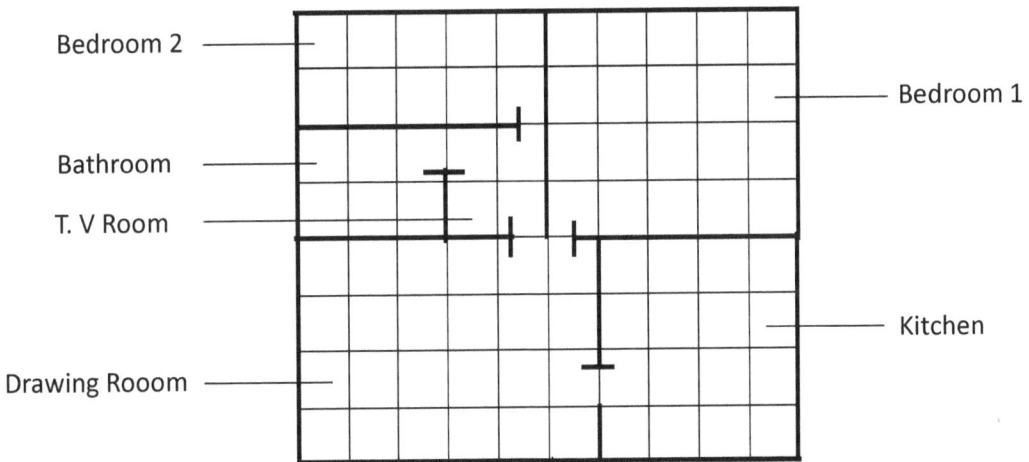

Bedroom 2

Bathroom

T. V Room

Drawing Rooom

Bedroom 1

Kitchen

Fig. Floor plan of a house

Observation

Room	Floor plan dimensions	Actual dimensions

Result

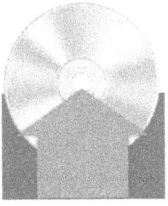

Two-Step Linear Equation

Aim
To solve $3x$

Materials required

Prerequisite Knowledge

Procedure

$x - 3 = 6$ by using the

Split them into equal groups.

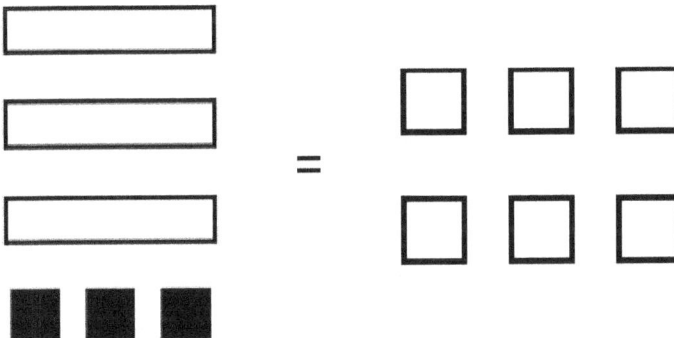

Fig:

Observation

	Algebraic/Numeric	
Right hand side		
groups of 3		

Result

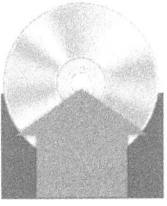

Project - 6

Tree Diagram of Probability

Aim

Materials required

Prerequisite Knowledge

$$Probability\,of\,an\,event = \frac{Number\,of\,favourable\,outcome}{Total\,number\,of\,outcome}$$

Procedure

from glazed paper.

Draw a boy or a girl on top of the drawing sheet.

$\frac{1}{2}$ $\frac{1}{2}$

Fig: A Tree diagram showing the probability of wearing one of the two pants

Draw a tree diagram showing his/her probability of wearing one of the two pants.

Extend the tree by showing his/her probability of wearing one of the three shirts with one of the two pants.

Further extend the tree by showing the shirts and pants with one of the two pairs of shoes.

Find the sum of probability of an event.

Observation

Event	Probability

Result

The sum of the probability of any event is _____ .

Clock Face Angles

Aim

Materials required

A pair of scissors, thin cardboard, ruler, pencil, compass, two straws, glazed paper, sketch pen, thumb pin.

Prerequisite Knowledge

° ° is known as a

Procedure

Make a clock face on the cardboard

Join the twelve hours with the

°).

Make the hands with the straw and

Find the   by using the
formula:

6, 3, 2:55, 3:45.

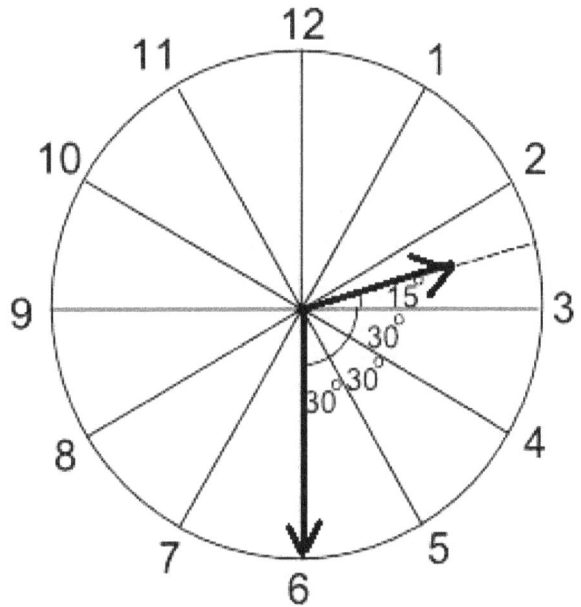

Fig: Clock face angles

Observation

Given Time	Angle (in degrees)	

Result

Area of a Rhombus

Aim
To derive the formula of the area of a rhombus.

Materials required
A pair of scissors, thin cardboard, ruler, pencil, glazed paper.

Prerequisite Knowledge
    – A parallelogram with all sides equal is known as a *rhombus*.

Area of a rectangle = Length × Breadth

Procedure

rhombi

Cut along the diagonals to get four right triangles from each rhombus.

Arrange the eight right triangles to get a rectangle.

Find the area of the rhombus using the rectangle.

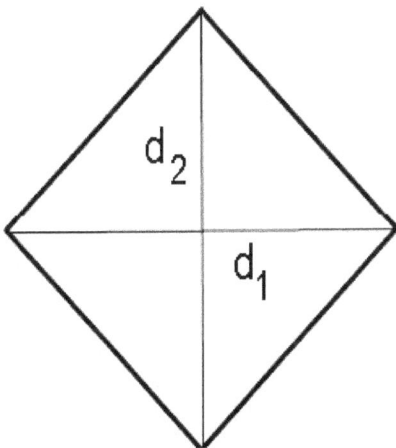

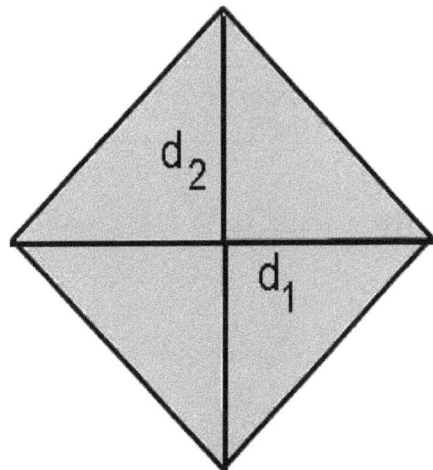

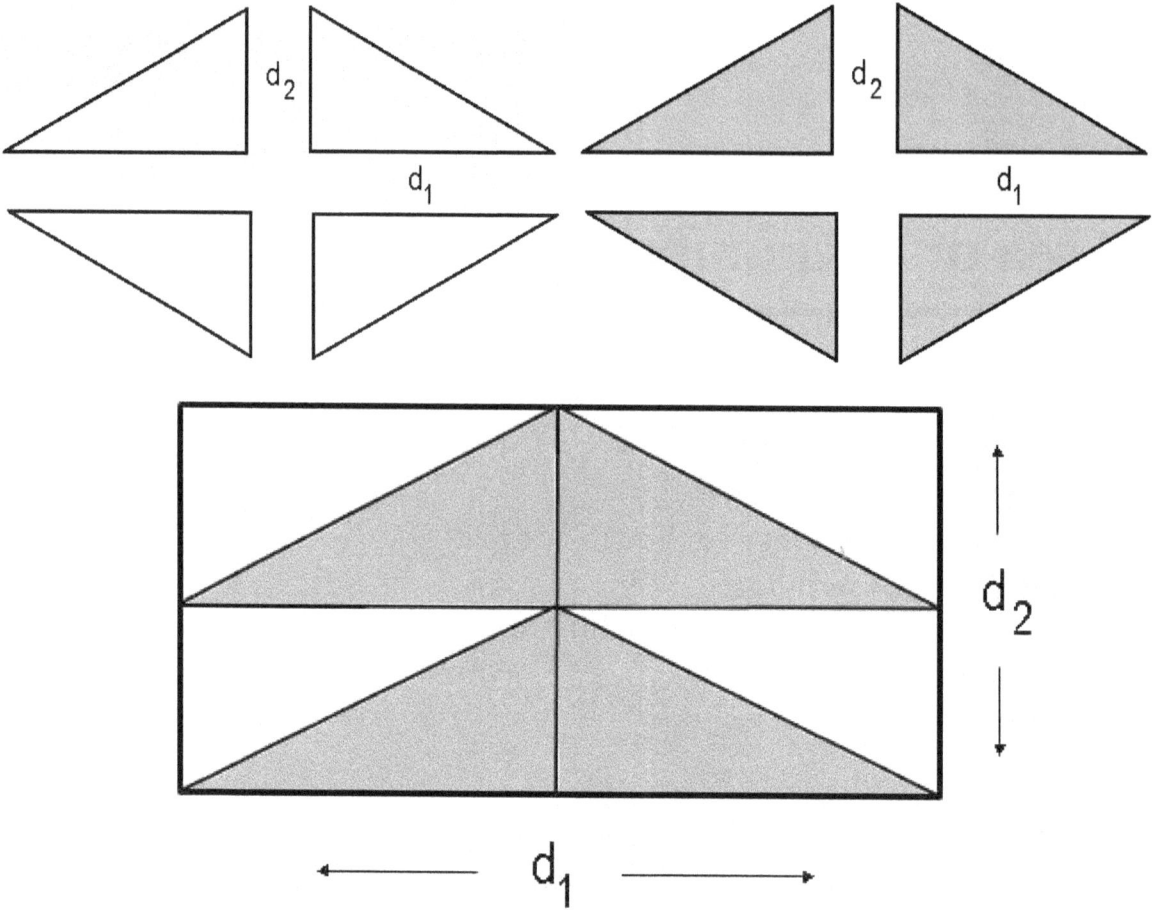

Fig:

Observation

The rectangle has _____ rhombi.

Diagonal d_1 corresponds to the _____ of the rectangle and the diagonal d_2 corresponds to the _____ of the rectangle.

Result

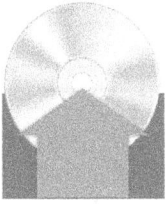

Surface Area of a Cuboid

Aim

curved surface area and the total *surface area* of a cuboid.

Materials required

Thin cardboard, a pair of scissors, glue, tracing or carbon paper, pencil.

Prerequisite Knowledge

***Curved Surface Area* —**
known as the *curved surface area of a cuboid*.

—

the

Area of a rectangle = Length × Breadth

Procedure

Trace or draw the net of the cuboid on a thin cardboard.

Cut it out along the boundary.

Fold it along the lines and glue the tabs to form a cuboid.

Find the area of the faces 1, 2, 3 and 4.

Find the area of the

face.

table.

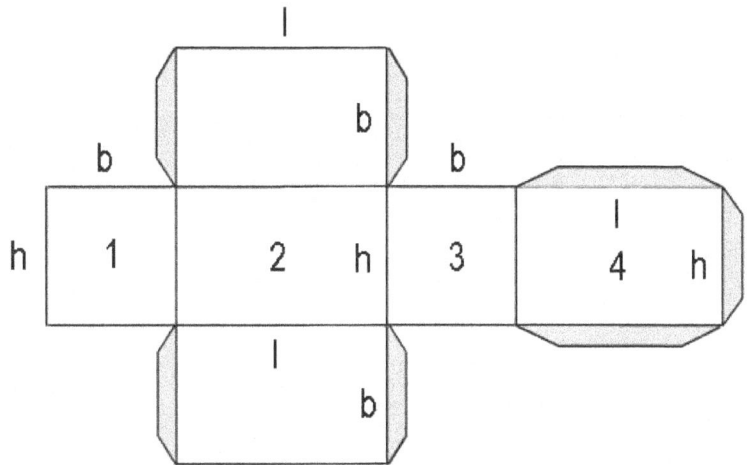

Fig: Net of a cuboid

Observation

Face	Area (in square units)
1	
2	
3	
4	
Top	

= Area of face 1 + Area of face 2 + Area of face 3 + Area of face 4

= _____.

= _____.

Result

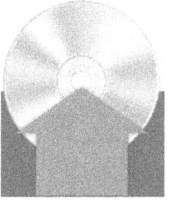

Spatial View of 3D Shapes

Aim

top, front and   of 3D shapes.

Materials required

Prerequisite Knowledge

– An object having
height) and no thickness.

An object having all the three dimensions, i.e.,   and height.

Procedure

Arrange the cubes in the given shapes on a table.

Stand and see the top view.

Next see the front view and the side view at the eye level.

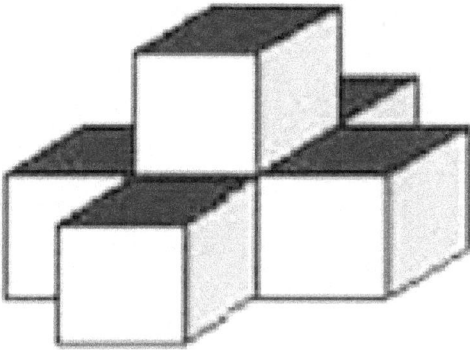

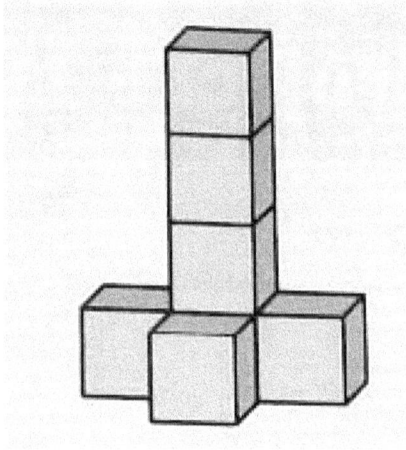

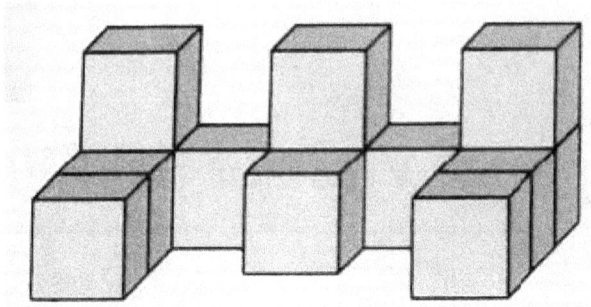

Fig:

Observation

Shape	Top View	Front View	Side View

Result

1. Distributive Property

Observation

Number		
36		3 × 12
21 + 15		3 × 7 + 3 × 5
21 + 15		3 × (7 + 5)

Result

36 = 21 + 15 = 3 × 7 + 3 × 5 = 3 × (7 + 5) = 3 × 12

2. Integers

Observation

Integers	First integer		Sum
+4 and +6			
+5 and -9			(Five zero pairs are removed)
-7 and - 3			

Result

+4 +6 = +10, +5- 9 = - 4, -7 - 3 = -10

3. Prime and Composite Numbers

Observation

Number			Prime/
7	⬡⬡⬡⬡⬡⬡⬡	1 × 7	Prime
9	(dots)	1× 9 3 × 3	Composite
15	(dots) or (dots)	1 × 15 5 × 3 Or 3 × 5	Composite
16	(dots) or (dots)	1 × 16 2 × 8 Or 8 × 2 4 × 4	Composite
19	(dots)	1 × 19	Prime

Result

7 and 19 are *prime numbers*. 9, 15 and 16 are *composite numbers.*

4. Factors and Multiples

Observation

	1×16
	2×8
	4×4

	1×16
	2×16
	3×16
	4×16

Result

5. Prime Factorisation

Observation

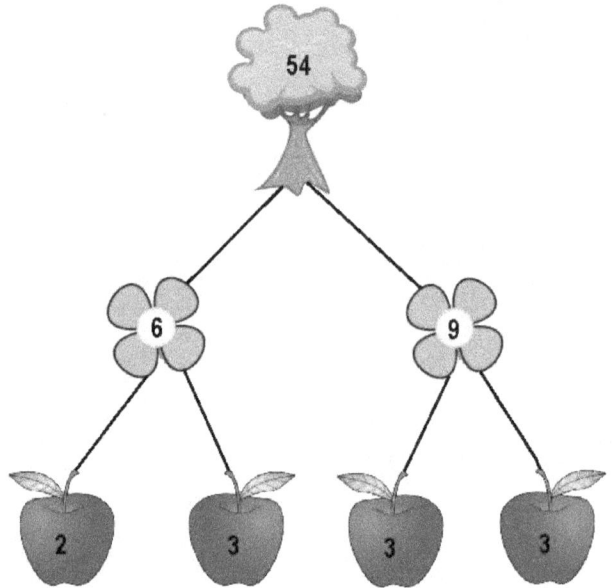

The prime factors of 30 are 2, 3 and 5.

The prime factors of 54 are 2, 3, 3 and 3.

Result

$30 = 2 \times 3 \times 5$ and $54 = 2 \times 3 \times 3 \times 3$

71+10 New Mathematics Projects

6. Highest Common Factor

Observation

32 cm and 12 cm		2	8 cm
12 cm and 8 cm		1	4 cm
8 cm and 4 cm		2	

Result

HCF of 32 and 12 is 4.

7. Improper and Mixed Fractions

Observation

3/2			$1\dfrac{1}{2}$
9/4			$2\dfrac{1}{4}$
10/3	Ten one-thirds		$3\dfrac{1}{3}$

Result

then write the answer in the form of $Q\dfrac{R}{D}$

8. Comparing Fractions

Observation

3/4, 7/11	$\frac{1}{4}$ $\frac{1}{4}$ $\frac{1}{4}$ $\frac{1}{11}$ $\frac{1}{11}$ $\frac{1}{11}$ $\frac{1}{11}$ $\frac{1}{11}$ $\frac{1}{11}$ $\frac{1}{11}$	$\dfrac{3}{4}$ $\dfrac{7}{11}$
2/7, 5/12	$\frac{1}{7}$ $\frac{1}{7}$ $\frac{1}{12}$ $\frac{1}{12}$ $\frac{1}{12}$ $\frac{1}{12}$ $\frac{1}{12}$	$\dfrac{2}{7}$ $\dfrac{5}{12}$
7/8, 6/8	$\frac{1}{8}$ $\frac{1}{8}$ $\frac{1}{8}$ $\frac{1}{8}$ $\frac{1}{8}$ $\frac{1}{8}$ $\frac{1}{8}$ $\frac{1}{8}$ $\frac{1}{8}$ $\frac{1}{8}$ $\frac{1}{8}$ $\frac{1}{8}$ $\frac{1}{8}$	$\dfrac{7}{8}$ $\dfrac{6}{8}$
3/10, 1/3	$\frac{1}{10}$ $\frac{1}{10}$ $\frac{1}{10}$ $\frac{1}{3}$	$\dfrac{3}{10}$ $\dfrac{1}{3}$
1/2, 4/9	$\frac{1}{2}$ $\frac{1}{9}$ $\frac{1}{9}$ $\frac{1}{9}$ $\frac{1}{9}$	$\dfrac{1}{2}$ $\dfrac{4}{9}$
3/6, 5/6	$\frac{1}{6}$ $\frac{1}{6}$ $\frac{1}{6}$ $\frac{1}{6}$ $\frac{1}{6}$ $\frac{1}{6}$ $\frac{1}{6}$ $\frac{1}{6}$	$\dfrac{3}{6}$ $\dfrac{5}{6}$

Result

numerator

9. Addition of Fractions

Observation

7/8 + 1/4	$\dfrac{7}{8}$ = [$\frac{1}{8}$ $\frac{1}{8}$ $\frac{1}{8}$ $\frac{1}{8}$ $\frac{1}{8}$ $\frac{1}{8}$ $\frac{1}{8}$] $\dfrac{1}{4}$ = [$\frac{1}{4}$]	7/8 + 1/4 = 7/8 + 2/8 = 9/8 $1\dfrac{1}{8}$
8/10 + 1/2	$\dfrac{8}{10}$ = [$\frac{1}{10}$ $\frac{1}{10}$ $\frac{1}{10}$ $\frac{1}{10}$ $\frac{1}{10}$ $\frac{1}{10}$ $\frac{1}{10}$ $\frac{1}{10}$] $\dfrac{1}{2}$ = [$\frac{1}{2}$]	8/10 +1/2 = 8/10 + 5/10 = 13/10 $1\dfrac{3}{10}$

Result

a) 7/8 + 1/4

  $1\dfrac{1}{8}$

  $1\dfrac{3}{10}$

10. Multiplication of Fractions

Observation

1/5 of 3/4	$\dfrac{1}{5}$	$\dfrac{3}{4}$	$\dfrac{3}{20}$
5/6 of 7/8	$\dfrac{5}{6}$	$\dfrac{7}{8}$	$\dfrac{35}{48}$

Result

$\dfrac{1}{5}$ of $\dfrac{3}{4}$ $\quad$ $\dfrac{1}{5}$ x $\dfrac{3}{4}$ $\quad$ $\dfrac{3}{20}$

$\dfrac{5}{6}$ of $\dfrac{7}{8}$ $\quad$ $\dfrac{5}{6}$ x $\dfrac{7}{8}$ $\quad$ $\dfrac{35}{48}$

11. Decimal Representation

Observation

0.8	
0.009	
0.36	
1.2	

71+10 New Mathematics Projects

2.29	  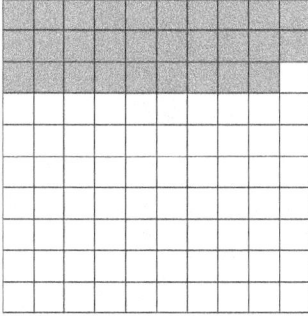
3.023	   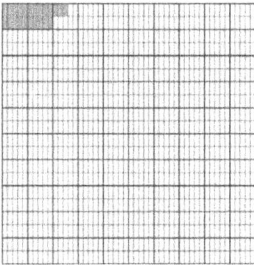

Result

12. Decimal Equivalent

Observation

0.4	
0.40	
0.400	
0.6	

0.60	
0.600	

0.4, 0.40 and 0.400 represent the *same*

0.6, 0.60 and 0.600 represent the *same*

Result

13. Addition of Decimals

Observation

First addend	
Second addend	
	4.78

First addend	
Second addend	
	 3.45

Result
1.1 + 3.68 = 4.78 and 2.36 + 1.09 = 3.45

14. Measurement

Observation

Result

15. Exponents

Observation

			2^0
	2		2^1
2	4		2^2

			2^3
4			2^4
5			2^5

Result

16. Square Root

Observation

		squares	
		4	2

	25	5
	49	7
	100	10

Result

17. Percentage

Observation

Design	Number squares	Number			
	25	5	$\dfrac{1}{5}$	0.2	20
	100	34	$\dfrac{34}{100}$	0.34	34
	100	38	$\dfrac{38}{100}$	0.38	38

Result

18. Ratio

Observation

Result

19. Proportion

Observation

Result

20. Variation

Observation

Triangle			Perpendicular / Base
ABC	18	12	1.5
	13.5	9	1.5
GHI	9	6	1.5
JKL	4.5	3	1.5

Result

21. Magic Square

Observation

Result

$$14 \times 3 = 42.$$

22. Coding Decoding

Observation

There is a *logical method* of   .

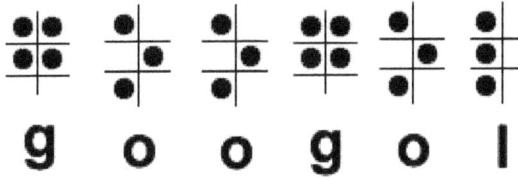

g o o g o l

Result

code language

23. Candle Clock

Observation

Result

24. Representing Polynomials

Observation

$^2 + 3$		

2		
2		

Result

25. Representing $(3a)^2$ and $3a^2$

Observation

a^2		a is a^2.
$3a^2$		$3a^2$ = Area of three side a.
$(3a)^2$		$(3a)^2$ side 3a = $9a^2$ = Area of nine side a.

Result

$3a^2$　　　　　　　　　　　　　　　　a, whereas $(3a)^2$ denotes the area of nine
　　　　　　a.

26. Adding Polynomials

Observation

2	
2	
2 2	

Result

2   2   2

27. Distributive Property

Observation

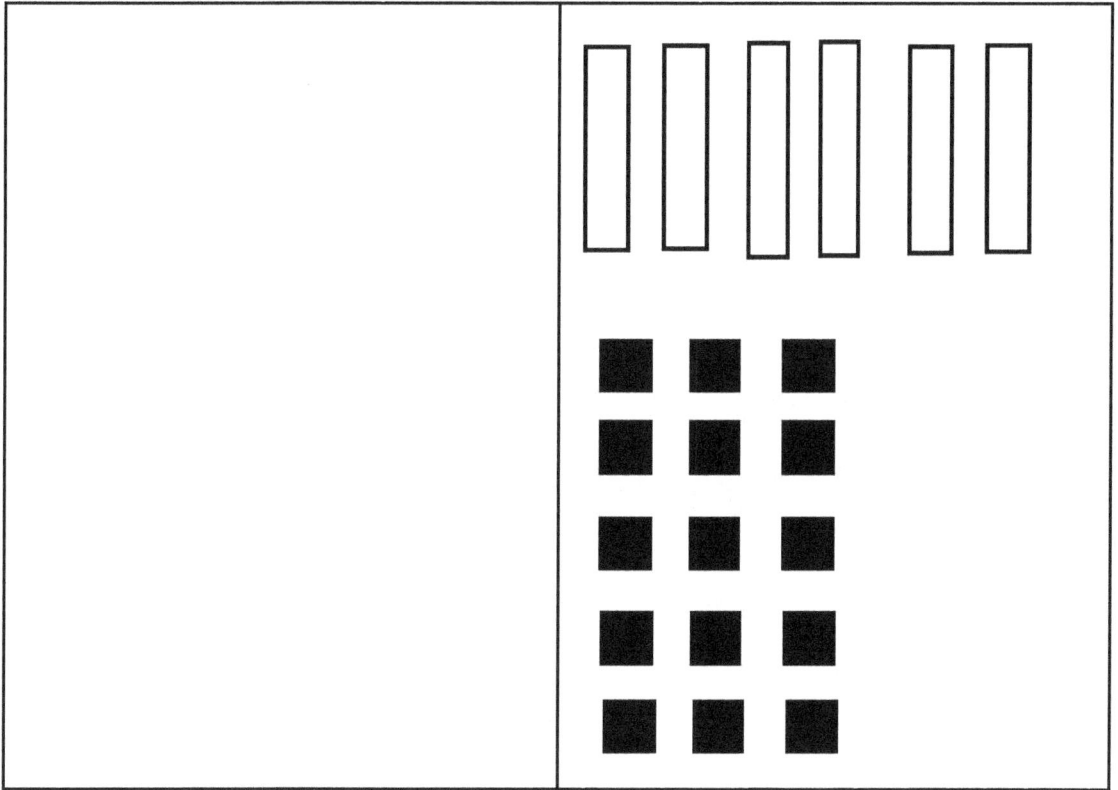

Result

28. Algebra Identity: $(a + b)^2 = a^2 + 2ab + b^2$

Observation

	a^2	2
	2	2
	2	

Result

^{2}

$= a^2$ ^{2}
$= a^2$ ^{2}
  $^2 = a^2$ ^{2}

29. Algebra Identity: $(a - b)^2 = a^2 - 2ab + b^2$

Observation

	2	2
	2	2
	a^2	

Result

$= a^2$

^{2} ^{2}
^{2} ^{2} ^{2} ^{2}
^{2} ^{2}
  $^2 = a^2$ ^{2}

30. Difference of Squares

Observation

a	a^2
	2

Result

$$a^2 - b^2$$

as $(a \qquad b \qquad a^2 - b^2 = (a \qquad b)$

31. Factorisation

Observation

	2	2
	10	

Result

$(x + 5)(x + 2)$

$= x^2 + 5x + 2x + 10$

$= x^2 + 7x + 10$

32. Addition of Linear Equation

Observation

		15	
		14	

Result

x = 9 and *y*

33. Subtraction of Linear Equation

Observation

	11	

zero pair		
side	11+3	

Result

x $x = 14$

34. Coordinates

Observation

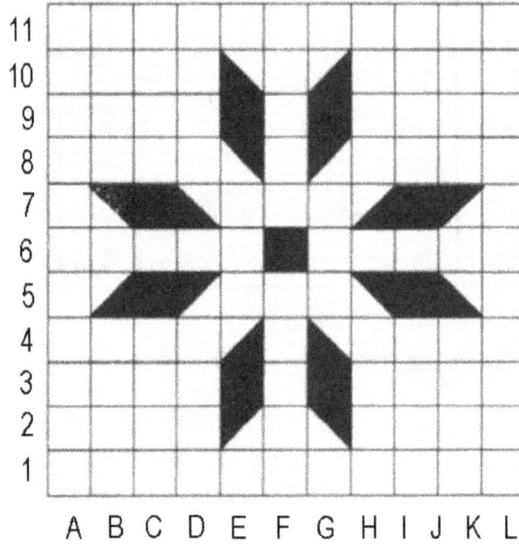

Fig:

Result

35. Enlargement of a Picture

Observation

(3,3), (0,6), (6,6), (6,9), (6,15), (15,9), (6,9) , (6,6), (15,6), (12,3), (3,3).

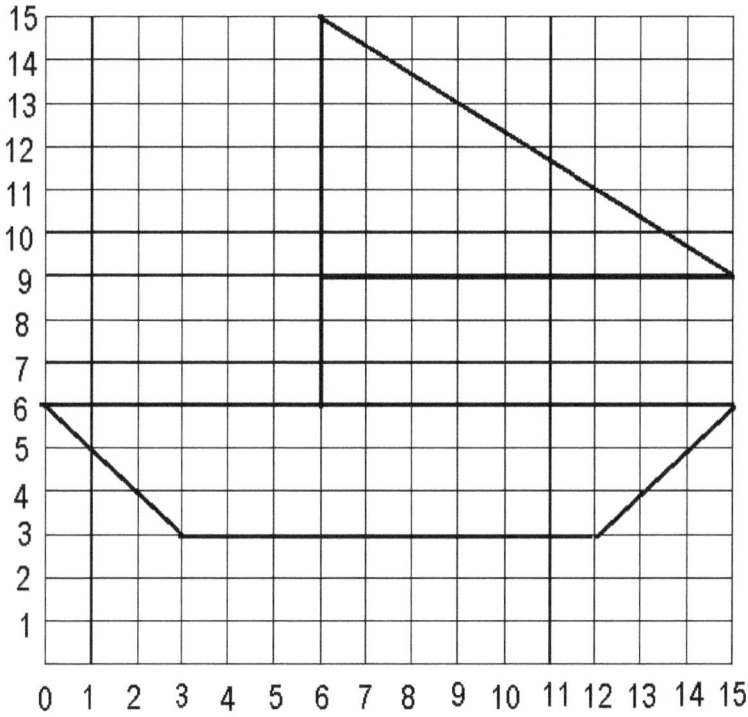

Fig:

Result

36. String Art

Observation

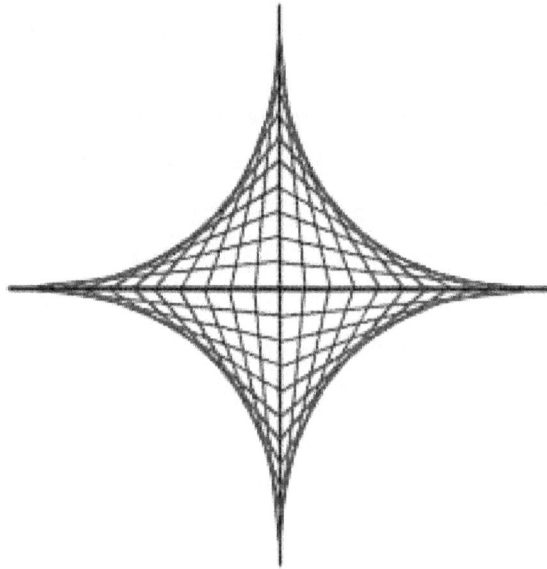

Fig:

Result

37. Line Graph

Observation

Result

38. Bar Graph

Result

39. Double Bar Graph

Observation

The

Result

40. Pie Chart

Observation

Result

41. Histogram

Observation

Result

42. Probability

Observation

Probability of tie $\dfrac{3}{9}$ $\dfrac{1}{3}$

Result

71+10 New Mathematics Projects

43. Mean of Numbers

Observation

Number		rearrangement	Mean = $\dfrac{\text{Total blocks}}{\text{Number of stacks}}$
5	35	7	35/5 = 7
6	36	6	36/6 = 6

$$\text{Mean} \quad \frac{\text{Sum of the terms}}{\text{Number of terms}}$$

Result

The **mean**

44. Tangram

Observation

		Two
		One
		One
		One
		Two

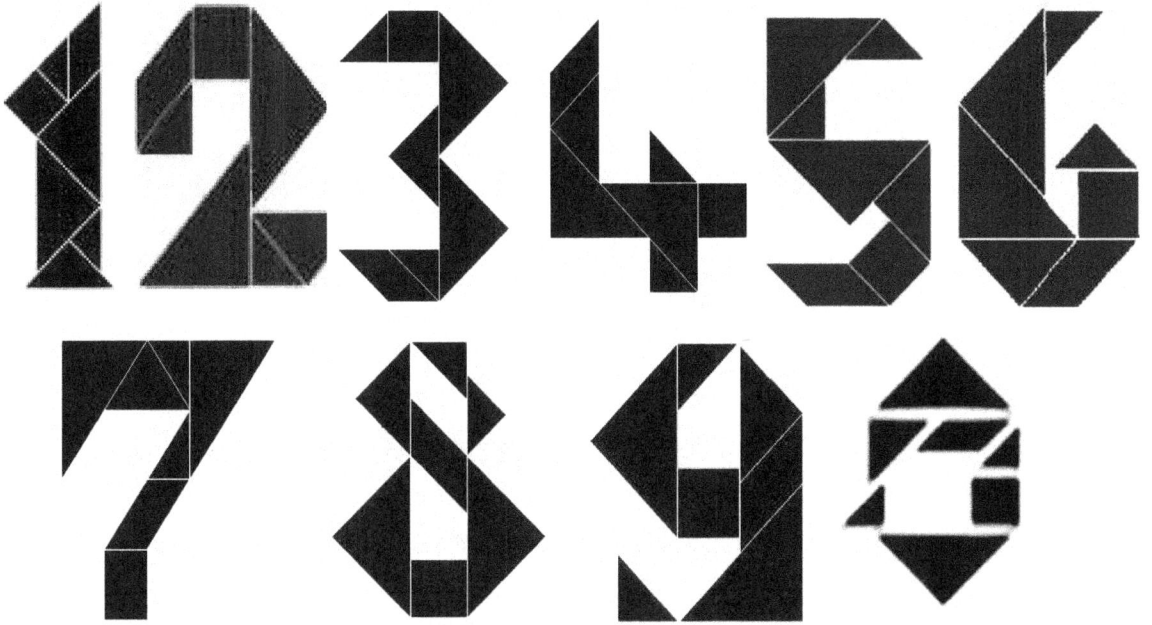

Fig:

Result

tangram set.

45. Line and Line Segment

Observation

Result

46. Angles formed by Parallel Lines

Observation

The *measurement of the angles*   *transversal* is drawn.

Corresponding angles:

$$\angle 1 = \angle 5 \qquad \angle 3 = \angle 7$$
$$\angle 2 = \angle 6 \qquad \angle 4 = \angle 8$$

Vertically opposite angles:

$$\angle 1 = \angle 4 \qquad \angle 5 = \angle 8$$
$$\angle 2 = \angle 3 \qquad \angle 6 = \angle 7$$

Interior angles on the same side of the transversal:

$$\angle 4 + \angle 6 = 180° \quad \angle 3 + \angle 5 = 180°$$

Result

opposite angles are equal and the *alternate interior angles on the same side of the transversal* are *supplementary.*

47. Angles on Magnetic Compass

Observation

Serial number		
1.	N and W	
2.		
3.		
4.		
5.		
6.	W and SW	

Result

are formed in a

48. Types of Triangles

Observation

Cases	triangles		angles	
Case I				
Case II				
Case III				
Case IV				

Result

acute

angled, obtuse angled or right angled.

49. Angle Sum Property of a Triangle

Observation

straight line.

180°.

$$\angle a + \angle b + \angle c = 180°$$

Result

The sum of the three angles of a triangle is 180°.   angle sum property of a triangle.

50. Sum of Length Property of a Triangle

Observation

6 cm, 8 cm and 15 cm		8 + 15 > 6, 6 + 15 > 8, 6 + 8 <15
6 cm, 8 cm and 14 cm		8 + 14 > 6, 6 + 14 > 8, 6 + 8 = 14
6 cm, 8 cm and 12 cm		12 + 6 > 8, 12 + 8 > 6, 6 + 8 >12

Result

51. Median of a Triangle

Observation

same point (points)
interior of the *triangle*

Result

centroid) of an *equilateral triangle*
the *interior* of the *triangle*.

52. Area of Congruent Triangles

Observation

ABC	8
	8
KLM	16
	16

congruent and have the *same area*.

PQR are not congruent,   same area.

Result

Two congruent triangles   *same area,*   *two triangles having the same area not be congruent.*

53. Pythagoras Theorem

Observation

	$25 = 5^2$
KLMN	$16 = 4^2$
	$9 = 3^2$

$25 = 16 + 9$

or, $5^2 = 4^2 + 3^2$

or, $AC^2 = AB^2 + BC^2$

Result

$^2 = (\text{Side}_1)^2 + (\text{Side}_2)^2$

54. Hexagon from a Circle

Observation

circle is a *hexagon*. It has *six sides,* *four sides.*

Result

A *hexagon*   *circle*   *rectangle.*

55. Angle Sum Property of a Quadrilateral

Observation

complete angle.

Angle around a point   *360°.*

$\angle a + \angle b + \angle c + \angle d = 360°$

Result

four angles of a quadrilateral is 360°.   *angle sum property* of a quadrilateral.

56. Area and Perimeter of a Square

Observation

Figure		unit squares	Area (in square		Perimeter
1	1	1	1	4	4
2	2	4	4	8	8
3	3	9	9	12	12
4	4	16	16	16	16
5	5	25	25	20	20

Result

2

57. Area and Perimeter of a Rectangle

Observation

Figure			unit squares	Area (in square		Perimeter
1	1	4	4	4	10	10
2	2	3	6	6	10	10
3	4	2	8	8	12	12
4	5	3	15	15	16	16

Result

The total number of unit squares = Area of a rectangle = Length × Breadth

The sum of four sides = Perimeter of a rectangle = 2(Length + Breadth)

58. Area of a Parallelogram

Observation

rectangle.

Result

Area of a rectangle = Length × Breadth

Area of a parallelogram = Base × Corresponding Height

59. Area of a Triangle

Observation

parallelogram.

Area of a parallelogram

Result

Area of a parallelogram = Base × Height

Area of a parallelogram

2 × Area of triangle = Base × Height

$$\text{or Area of triangle} = \frac{1}{2} Base \quad Height$$

60. Same Area, Different Perimeters

Observation

	2	
	36	24
	36	30

Result

61. Properties of Square and Rhombus

Observation

	Square	
	each other.	

Result

rhombus is a square   *every square is not a rhombus.*

62. Properties of Parallelogram and Rectangle

Observation

Length of opposite sides		

Result

Every parallelogram is a rectangle,   *every rectangle is not a parallelogram.*

63. Pi is a Constant

Observation

The *circumference* and the *diameter*   *size*

Result

64. Area of a Circle

Observation

Height   *Radius*   *r.*

Base of the parallelogram $= \dfrac{1}{2}$ x Circumference of the circle

Base of the parallelogram $= \dfrac{1}{2}$ x 2πr

Base of the parallelogram $= \pi r$

Result

Area of the circle

Area of the parallelogram ^{2}

^{2}

65. Surface Area of a Cube

Observation

1	a^2
2	a^2
3	a^2
4	a^2
Top	a^2
	a^2

= Area of face 1 + Area of face 2 + Area of face 3 + Area of face 4

$= a^2 + a^2 + a^2 + a^2 = 4\,a^2$

$= 4\,a^2 + a^2 + a^2 = 6\,a^2$

Result

Total surface area of a cube = 6 a

66. Animal Shape Using Unit Cubes

Observation

Result

67. Surface Area Comparison

Observation

	18
Two on top of the other two	16

Result

changes with the rearrangement of the cube, and *hence, the surface area also changes.*

68. Volume of Cube and Cuboid

Observation

Cube	number	Number			
	27	3	3	3	3 × 3 × 3 = 27

	number	Number	Number		
	30	5	3	2	5×3×2 = 30

Result

Volume of cube
Volume of cuboid

69. Area of a Cylinder

Observation

Length × Breadth

2   2

Result

70. 3D Shapes from the Net

Observation

		Number	Number	
		2	0	1
Cone		1	1	2
	5	8	5	2

Result

There are *3 faces,*   and *no vertex of a cylinder.* There are
cone. There are *5 faces, 8 edges* and

A

71. Symmetrical Design

Observation

four.

rhombi

Fig:

Result

A symmetrical lacy design is obtained from a circle.

10 New Projects

1. Square Numbers

Observation

		unit squares	
1		1	1
2		4	4 = 1 + 3 or 4 = 2 × 2
3		9	9 = 1 + 3 + 5 or 9 = 3 × 3
4		16	16 = 1 + 3 + 5 + 7 or 16 = 4 × 4
5		25	25 = 1 + 3 + 5 + 7 + 9 or 25 = 5 × 5

Result

71+10 New Mathematics Projects

2. Equivalent Fractions

Observation

1/4	$\frac{1}{4}$ $\frac{1}{8}$ | $\frac{1}{8}$ $\frac{1}{12}$ | $\frac{1}{12}$ | $\frac{1}{12}$	1/4 = 1/8 + 1/8 1/4 = 1/12 + 1/12 + 1/12
1/3	$\frac{1}{3}$ $\frac{1}{6}$ | $\frac{1}{6}$ $\frac{1}{9}$ | $\frac{1}{9}$ | $\frac{1}{9}$	1/3 = 1/6 + 1/6 1/3 = 1/9 + 1/9 + 1/9
1/2	$\frac{1}{2}$ $\frac{1}{4}$ | $\frac{1}{4}$ $\frac{1}{6}$ | $\frac{1}{6}$ | $\frac{1}{6}$ $\frac{1}{8}$ | $\frac{1}{8}$ | $\frac{1}{8}$ | $\frac{1}{8}$ $\frac{1}{10}$ | $\frac{1}{10}$ | $\frac{1}{10}$ | $\frac{1}{10}$ | $\frac{1}{10}$ $\frac{1}{12}$ | $\frac{1}{12}$ | $\frac{1}{12}$ | $\frac{1}{12}$ | $\frac{1}{12}$ | $\frac{1}{12}$	1/2 = 1/4 + 1/4 1/2 = 1/6 + 1/6 + 1/6 1/2 = 1/8 + 1/8 + 1/8 + 1/8 1/2 = 1/10 + 1/10 + 1/10 + 1/10 + 1/10 1/2 = 1/12 + 1/12 + 1/12 + 1/12 + 1/12 + 1/12
1/5	$\frac{1}{5}$ $\frac{1}{10}$ | $\frac{1}{10}$	1/5 = 1/10 + 1/10

3/4		3/4= 1/8 + 1/8 + 1/8 + 1/8 + 1/8 + 1/8 3/4 = 1/12 + 1/12 + 1/12 +1/12 + 1/12 + 1/12 + 1/12 + 1/12 + 1/12
2/5		2/5 = 1/10 + 1/10 + 1/10 + 1/10
1/6		1/6 = 1/12 + 1/12

Result

1/4	2/8 , 3/12
1/3	2/6 , 3/9
1/2	2/4 , 3/6, 4/8, 5/10, 6/12
1/5	2/10
3/4	6/8, 9/12
2/5	4/10
1/6	2/12

71+10 New Mathematics Projects

3. Multiplication of Decimals

Observation

0.8 × 0.3				0.24
0.4 × 0.7				0.28

Result

0.8 × 0.3 = 0.24 and 0.4 × 0.7 = 0.28

4. Scale Drawing

Observation

	L = 6, B = 4	L = 9, B = 6
Kitchen	L = 4, B = 4	L = 6, B = 6
Bedroom 1	L = 5, B = 4	L = 7.5, B = 6
Bedroom 2	L = 5, B = 2	L = 7.5, B = 3
Bathroom	L = 3, B = 2	L = 4.5, B = 3
TV room	L = 2, B = 2	L = 3, B = 3

Result

architects, geographers, etc.   *scale drawing*

5. Two-Step Linear Equation

Observation

	6	
zero pair		
side	6 + 3	

side		
	9 (3 + 3 + 3)	

Result

x   *x*   *x* = 3.

6. Tree Diagram of Probability

Observation

Wearing either pair of pant	1/2
	1/6
	1/12

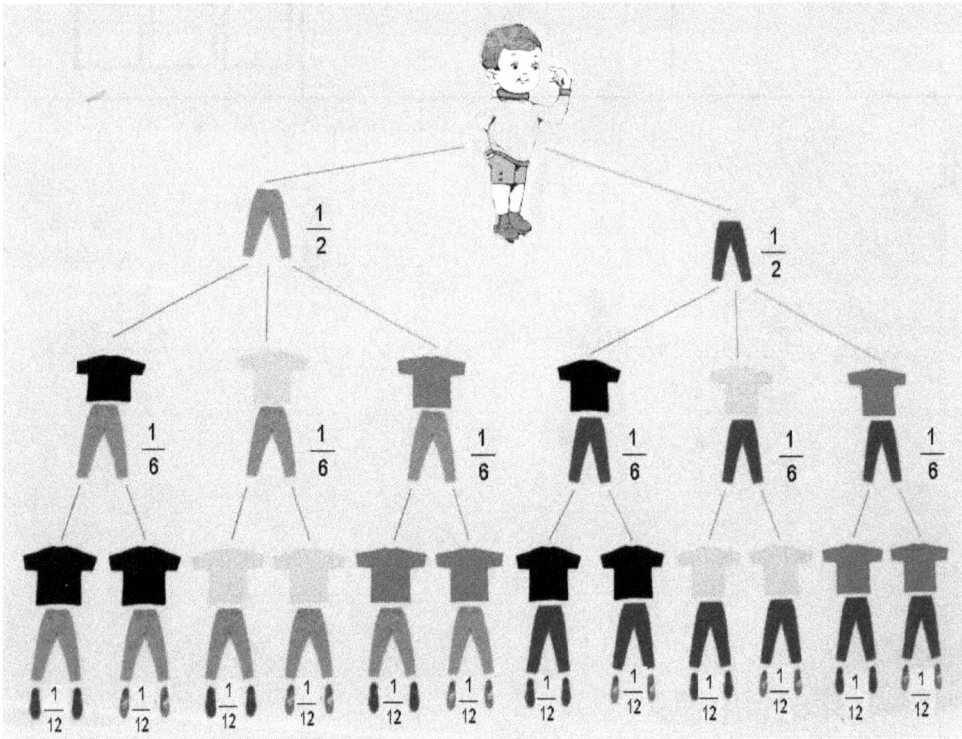

Fig:

Result

7. Clock Face Angles

Observation

Time	Angle	
11		
6		
3		

Result

analog clock   *angle.*

8. Area of a Rhombus

Observation

two rhombi.

Diagonal d_1 corresponds to the length   *diagonal d   corresponds to the*
breadth

Result

Area of a rectangle = Length × Breadth = $d_1 × d_2$

Area of rectangle

$d_1 × d_2$ = 2 × Area of a rhombus

or Area of Rhombus $= \dfrac{1}{2} d_1 \ d_2$

9. Surface Area of a Cuboid

Observation

1	Breadth × Height
2	Length × Height
3	Breadth × Height
4	Length × Height
Top	Length × Breadth
	Length × Breadth

= Area of face 1 + Area of face 2 + Area of face 3 + Area of face 4

Result

10. Spatial View of 3D Shapes

Observation

Result

GREATEST CRAFTS PROJECTS for CHILDREN

WONDERS of the World

RAMAYANA

NEW Learning MATHEMATICS The Fun Way

71 ARTS & CRAFTS FOR SCHOOL CHILDREN

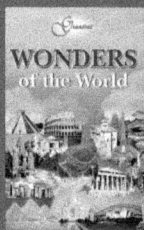

Also Available in Hindi

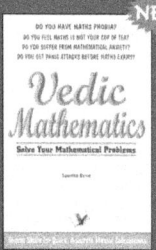

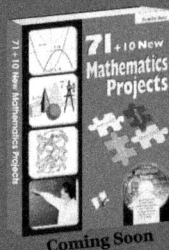

NEW Vedic Mathematics Solve Your Mathematical Problems

वैदिक गणित

Amazing Science Activities

Also Available in Hindi

71 +10 New Mathematics Projects

Also Available in Hindi

71 +10 New Science Projects Junior

Also Available in Hindi

Also Available in Hindi

71 +10 New Mathematics Projects

Coming Soon

Also Available in Hindi, Tamil & Bangla

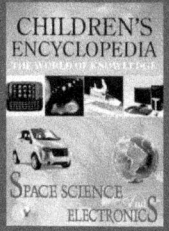

Drawing & Painting Course

CHILDREN'S ENCYCLOPEDIA THE WORLD OF KNOWLEDGE

CHILDREN'S ENCYCLOPEDIA THE WORLD OF KNOWLEDGE GENERAL KNOWLEDGE

CHILDREN'S ENCYCLOPEDIA THE WORLD OF KNOWLEDGE LIFE SCIENCES and HUMAN BODY

CHILDREN'S ENCYCLOPEDIA THE WORLD OF KNOWLEDGE PHYSICS and CHEMISTRY

CHILDREN'S ENCYCLOPEDIA THE WORLD OF KNOWLEDGE SCIENTIFIC INVENTIONS and DISCOVERIES

CHILDREN'S ENCYCLOPEDIA THE WORLD OF KNOWLEDGE SPACE SCIENCE and ELECTRONICS

Contact us at sales@vspublishers.com

HINDI LITERATURE

TALES & STORIES

MUSIC (संगीत)

MYSTERIES (रहस्य)

MAGIC & FACT (जादू एवं तथ्य)

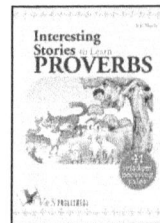

CHILDREN TALES (बच्चों की कहानियाँ)

All books available at www.vspublishers.com

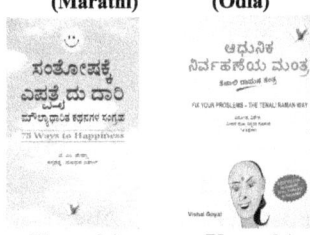

www.ingramcontent.com/pod-product-compliance
Lightning Source LLC
Chambersburg PA
CBHW080527220326
41599CB00032B/6229